SCIENCE

For the HiSET® Test

Photos courtesy of:

p. 17: © arindambanerjee; p. 40: © Taiga; p. 49: © dorsetman; p. 74: © Razvan; p. 106: © obalt Moon Design; p. 132: © KPG Payless2; p. 158: © demarfa

Science for the HiSET® Test
ISBN 978-1-56420-883-5

New Readers Press
ProLiteracy's Publishing Division
104 Marcellus Street, Syracuse, New York 13204
www.newreaderspress.com

Printed in the United States of America
10 9 8 7 6 5 4 3 2 1

Proceeds from the sale of New Readers Press materials support professional development, training, and technical assistance programs of ProLiteracy that benefit local literacy programs in the U.S. and around the globe.

Developer: QuaraCORE
Editorial Director: Terrie Lipke
Cover Design: Carolyn Wallace
Technology Specialist: Maryellen Casey

Contents

Using This Book

Welcome to *Science for the HiSET® Test*, an important resource in helping you build a solid foundation of science skills as you prepare for the HiSET® high school equivalency test.

- First, read *What to Expect* on page 5, which will give you a brief overview of the HiSET® test itself.
- Next, take the Pretest, which begins on page 6. After taking the Pretest and checking your answers, use the chart on page 16 to find the lessons that will help you study the skills you need to improve.
- Then, start using the book, which is organized into four units, each containing brief lessons that focus on specific themes and skills. Important vocabulary terms included in the unit are listed on the first page of the unit and appear in **boldface** when they are first used in each lesson. Use the Glossary at the end of each unit to find key word definitions.
- Each lesson is followed by a Lesson Practice with questions to test your knowledge of the lesson content. Answers can be found in the Answer Key at the end of the unit. Each Lesson Practice also includes *Key Point!* and *Test Strategy* tips to help you prepare for the HiSET® test.
- Each unit concludes with a Unit Test that covers all the content in the unit's lessons. The Unit Test Answer Key appears at the end of each unit.
- Every unit concludes with *Study More!*, which lists additional skills and topics you can study to prepare for the HiSET® test.
- After completing all the units, you can test what you know by taking the HiSET® Practice Test, beginning on page 179. This test will help you check your understanding of all the skills in the book.

What to Expect

This book is intended to help you prepare to take the HiSET® (High School Equivalency Test) Exam in Science. This is one of the five HiSET® exams; the others are in Mathematics, Reading, Writing, and Social Studies. The HiSET® exams are available in English and in Spanish and can be taken in written format (on paper) or on a computer. For more information about the HiSET® exams, visit hiset.ets.org/test_takers.

Preparing for the Test

Using, *Science for the HiSET® Test*, is a great first step in preparing to take the test. This book provides an overview of the content of the Science test and gives you many opportunities to take practice tests to evaluate how well you know the content and skills that will be included in the HiSET® test.

Be sure to allow enough time to prepare for the test. Choose a test date that will not force you to rush through your study period. You need time to use this book, and possibly more time for additional review and practice tests.

When using this book, the Pretest helps you identify your strengths and weaknesses in science content and skills so you can pinpoint the areas on which to focus your study for the HiSET®.

Taking the Science Test

The HiSET® Science exam includes 50 multiple-choice questions. You will have 80 minutes to complete the test.

The Science exam will assess your understanding of key content and skills taught in high school, including:

- the scientific process
- earth science
- space science
- physical science
- life science

These and many other topics are reviewed in this book. Each lesson covers a specific topic, and each is followed by practice questions to help you learn the content. After you complete the practice HiSET® test at the end of this book, you can judge your results and decide if you are ready to take the HiSET® Science test.

Answer the following questions to gauge your readiness to take the HiSET® test. Answers to questions are on pages 14–16.

1. Which is true of lizards and fish?
   - **A** They both have gills.
   - **B** They are both cold blooded.
   - **C** Lizards have scales; fish have a smooth skin.
   - **D** Lizards lay eggs; fish give birth to live young.

2. Which of the following would slow the rate at which nonrenewable resources are consumed?
   - **A** A factory installs a device on its smokestack to reduce the amount of pollution it emits.
   - **B** A restaurant begins composting its food waste to reduce the amount of garbage it sends to landfills.
   - **C** A power plant switches from natural gas fuel to wind power to reduce its impact on the environment.
   - **D** A fast food company switches from paper cups to plastic cups to reduce the number of trees used to produce paper.

3. Why are the results of an investigation with a large number of test subjects more reliable than one with very few test subjects?
   - **A** It eliminates the risk of bias.
   - **B** It helps reduce the limitations.
   - **C** It is easier to control the variables.
   - **D** It makes the hypothesis falsifiable.

4. Which pair of elements would be most likely to form a covalent bond?
   - **A** lithium (Li) and chlorine (Cl)
   - **B** chlorine (Cl) and bromine (Br)
   - **C** potassium (K) and lithium (Li)
   - **D** bromine (Br) and potassium (K)

5. Which is produced during the process of meiosis?
   - **A** muscle cells
   - **B** nerve cells
   - **C** sperm cells
   - **D** white blood cells

6. A soccer player kicks a 0.5 kg ball, giving it an acceleration of 18 m/s^2. How much force was in the kick?
   - **A** 4.5 N
   - **B** 9 N
   - **C** 18.5 N
   - **D** 36 N

Question 7 refers to the following image.

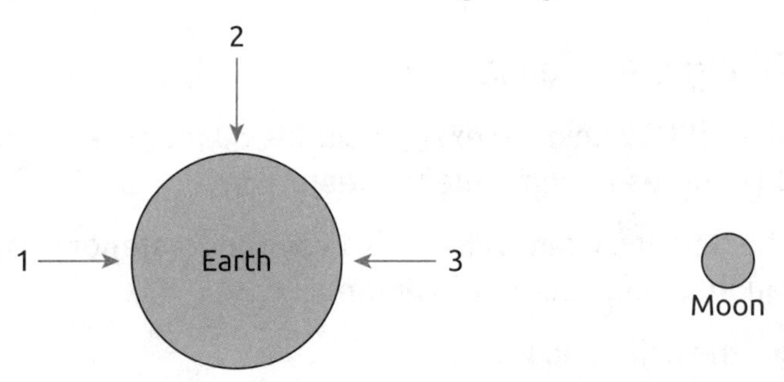

7. The diagram shows the position of the moon in relation to Earth. The numbers refer to different places on Earth's surface. Which point or points would experience low tide?
   - **A** 1
   - **B** 2
   - **C** 3
   - **D** 1 and 3

8. Which of the following is a biotic factor?
   - **A** decomposer
   - **B** oxygen
   - **C** rock
   - **D** water

Questions 9–11 refer to the following information.

A researcher placed 10 female fruit flies and 10 male fruit flies in the same vial. The vial contained 1 gram of fruit fly food. She placed the vial in an incubator at 23.9 degrees Celsius for 3 hours. She recorded how many males and how many females survived at the end of the 3 hours. She then repeated the experiment at 29.4 degrees Celsius, 26.7 degrees Celsius, 21.1 degrees Celsius, and 18.3 degrees Celsius.

9. What was the dependent variable in this investigation?
   A the temperature of the incubator
   B the ratio of male to female fruit flies at the start
   C the amount of time the fruit flies were kept in the incubator
   D the number of male and female fruit flies after 3 hours

10. What was the independent variable in this investigation?
   A the type of fruit fly food used
   B the temperature of the incubator
   C the amount of time the fruit flies were observed
   D the ratio of male to female fruit flies after 3 hours

11. Which statement would be an observation?
   A Female fruit flies are better able to adapt to changing temperature conditions.
   B If the temperature deviates greatly from room temperature, then more female fruit flies will survive.
   C More male fruit flies were living at the end of 3 hours when the temperature was 21.1 degrees Celsius.
   D More male fruit flies will be found living inside homes than outside, where there are more fluctuations in temperature.

12. An astronomer is studying two dying stars. Star A is a red supergiant, and Star B is a red giant. What can the astronomer predict about the future of the two stars?
   A Star A will one day become a neutron star, and Star B will become a black hole.
   B Star B will eventually become a red supergiant like Star A.
   C Star A will eventually explode in a supernova, and Star B will become a white dwarf.
   D Both stars will eventually become white dwarfs, then black dwarfs.

Questions 13 and 14 refer to the following image.

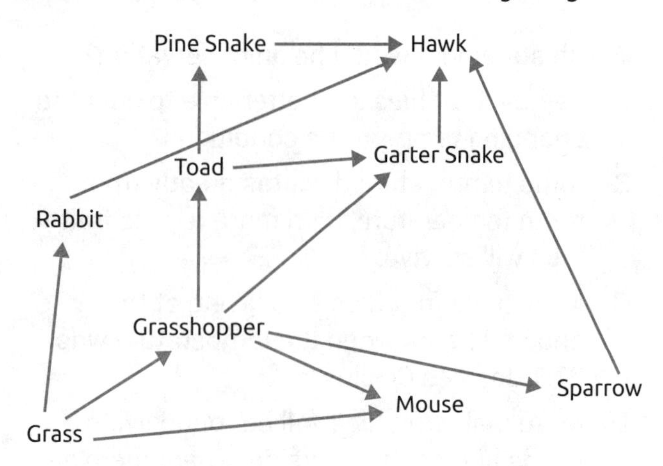

13. Which organism would most likely increase in population if the toad were taken out of the environment?
    A rabbit
    B grass
    C pine snake
    D grasshopper

14. Which organism is a producer?
    A grass
    B mouse
    C rabbit
    D grasshopper

15. Which scenario describes the most work being done?
    A A person pushes on a wall with 50 N of force.
    B A person pulls on a locked door with 50 N of force.
    C A person pushes on a 5 kg rock with 30 N of force for 1 meter.
    D A person pushes on a 5 kg rock with 20 N of force for 2 meters.

Question 16 refers to the following information.

Scientists use weather balloons to study conditions in Earth's atmosphere. The balloons carry instruments to measure and record weather data, such as temperature. When a balloon reaches a particular height, it bursts, and the instruments parachute back to earth, where scientists can retrieve the recorded data.

At 9:00 one morning, a scientist launches a weather balloon, and the balloon ascends steadily until it bursts at 10:10 a.m. When the instruments are recovered, the scientist finds that the temperature readings decreased between 9:00 and 9:40. Then, from 9:40 until the balloon burst at 10:10, the temperature readings increased.

16. In which layer of the atmosphere was the balloon located when it burst?
    A mesosphere
    B stratosphere
    C thermosphere
    D troposphere

17. The scientific name for the brown bear is *Ursus arctos*. The word *Ursus* is the bear's _______.
    - **A** class
    - **B** genus
    - **C** kingdom
    - **D** species

18. Carbon-14 is a substance found in the tissues of all living things. After an organism dies, the amount of carbon-14 in its tissues declines at a predictable rate. This allows scientists to use the amount of carbon-14 in the preserved tissues of ancient organisms to estimate how long ago the organisms lived. Scientists could use carbon-14 dating to estimate the age of which of the following objects?
    - **A** a meteorite
    - **B** a bone from a woolly mammoth
    - **C** a dinosaur track preserved in stone
    - **D** a trace fossil of an extinct fish

19. Which is a result of the Law of Conservation of Matter in a chemical reaction?
    - **A** The mass of each reactant must not change.
    - **B** The number of reactants must equal the number of products.
    - **C** The mass of the products will always be less than the mass of the reactants.
    - **D** The number of each type of atom must be the same in the reactants as in the products.

20. In which structures of the circulatory system does gas exchange occur?
    - **A** alveoli
    - **B** arteries
    - **C** bronchi
    - **D** capillaries

21. Which correctly describes what happens as the wavelength of electromagnetic radiation increases?
    - **A** Frequency and energy both increase.
    - **B** Frequency and energy both decrease.
    - **C** Frequency increases, and energy decreases.
    - **D** Frequency decreases, and energy increases.

22. Plants require nitrogen to grow and reproduce. Most nitrogen in soil is not in a form that plants can use. Which process makes soil nitrogen usable by plants?
    - **A** Rainstorms deposit plant-usable nitrogen into the soil.
    - **B** A specific type of bacteria in the soil converts soil nitrogen to a plant-usable form.
    - **C** Leaching moves nitrogen deeper into the soil.
    - **D** Farmers till the soil, converting soil nitrogen to a plant-usable form.

23. Which describes a chemical change?
    - **A** Liquid water is heated, producing bubbles of water gas.
    - **B** Red ink is spilled on white paper, turning the paper a light red color.
    - **C** Solid mercuric oxide is heated, producing liquid mercury and oxygen gas.
    - **D** Sodium chloride is dissolved in water, and the temperature remains the same.

24. A baby is able to communicate using sign language. This is an example of a(n) _______.
    - **A** reflex
    - **B** instinct
    - **C** innate behavior
    - **D** learned behavior

Questions 25–28 refer to the following information and table.

A group of students have three antacid tablets, all the same size and makeup. They have three beakers full of water, labeled X, Y, and Z. They drop one tablet into each of the beakers and then time how long it takes for the tablet in each beaker to dissolve. Their results are shown in the table.

Sample	Temperature of Water (°C)	Time for Tablet to Dissolve (seconds)
X	20	40
Y	30	25
Z	40	10

25. Which of the following statements is an inference?
   A The tablet in sample Z dissolved the fastest.
   B The tablet in sample X was in the warmest water.
   C The water in sample Z must have been stirred.
   D Temperature affects the rate at which an antacid tablet will dissolve.

26. What was a weakness in this investigation?
   A The hypothesis was proven.
   B Only one trial was conducted.
   C Different water temperatures were used.
   D The tablets were dropped at the same time.

27. Which of the following could be a conclusion for this experiment?
   A The tablet in sample X dissolved the slowest.
   B When a solution containing antacid is stirred, it dissolves faster.
   C If water temperature is high, then an antacid will dissolve quickly.
   D The temperature of water affects the rate at which an antacid will dissolve.

28. How would the results of this experiment be considered reliable?
   A The hypothesis is proven.
   B The conclusion is without bias.
   C More trials are added to the experiment.
   D Another group of students gets similar results.

29. Which layers of Earth are liquid?
   A outer core and inner core
   B lithosphere and outer core
   C inner core and asthenosphere
   D crust and inner core

Questions 30–32 refer to the following information.

A chemistry student dissolves a small amount of potassium hydroxide in 1 liter of water, making a clear solution. She measures the pH of the water before mixing in the potassium hydroxide and then measures the pH of the solution after. The pH of the water is 7.1. The pH of the solution is 11.2. She adds more potassium hydroxide to the solution and stirs. The pH of the new solution is 13.3.

30. Which best describes potassium hydroxide and the solution?
   - **A** Both potassium hydroxide and the solution are compounds.
   - **B** Both potassium hydroxide and the solution are mixtures.
   - **C** Potassium hydroxide is an element, and the solution is a mixture.
   - **D** Potassium hydroxide is a compound, and the solution is a mixture.

31. The pH change of the water when the potassium hydroxide was added indicates that potassium hydroxide is a(n) _______.
   - **A** base
   - **B** acid
   - **C** neutral substance
   - **D** nonionic substance

32. Which best describes potassium hydroxide and the water in this experiment?
   - **A** Potassium hydroxide is the solvent, and water is the solute.
   - **B** Water is the solution, and potassium hydroxide is the solute.
   - **C** Water is the solvent, and potassium hydroxide is the solute.
   - **D** Potassium hydroxide is the solution, and water is the solute.

33. An earthquake under the ocean can cause which of these natural disasters?
   - **A** typhoon
   - **B** hurricane
   - **C** tsunami
   - **D** supercell

34. A ball falls off a table toward the ground. Which describes the relationship between the kinetic and potential energy in the ball?
   - **A** Both potential and kinetic energy are decreasing.
   - **B** Both potential and kinetic energy are increasing.
   - **C** Potential energy is decreasing, while kinetic energy is increasing.
   - **D** Kinetic energy is decreasing, while potential energy is increasing.

35. If a mass of polar air moves into an area, replacing a mass of tropical air, what would the boundary where the two air masses meet be called?
   - **A** a warm front
   - **B** a cold front
   - **C** a polar front
   - **D** a tropical front

36. A car is traveling at a constant forward velocity of 20 m/s. Which situation involves the greatest acceleration?
   - **A** After 2 seconds, the car has a velocity of 25 m/s.
   - **B** After 3 seconds, the car has a velocity of 30 m/s.
   - **C** After 4 seconds, the car has a velocity of 20 m/s.
   - **D** After 5 seconds, the car has a velocity of 35 m/s.

37. Which of the following features do Earth and the sun have in common?
    - **A** Both are made from a mixture of solid and liquid rock.
    - **B** Both have gravitational fields of about the same size.
    - **C** Both are about the same temperature.
    - **D** Both have atmospheres.

38. Which best explains the function of the phloem vessels?
    - **A** transporting water from roots to leaves
    - **B** transporting sugar from leaves to roots
    - **C** transporting water from leaves to roots
    - **D** transporting sugar from roots to leaves

39. Why do scientists repeat work done by other scientists?
    - **A** to verify results
    - **B** to dispute a hypothesis
    - **C** to change the variables
    - **D** to make new observations

40. Which best describes a material that is made entirely of the same type of atom, is shiny and brittle, and conducts electricity?
    - **A** metal
    - **B** nonmetal
    - **C** metalloid
    - **D** compound

41. If Earth's axis were not tilted, which of the following changes would we experience?
    - **A** There would be no seasons.
    - **B** There would be no day and night.
    - **C** There would be no phases of the moon.
    - **D** There would be no solar eclipses.

42. Where are voluntary muscles found?
    - **A** esophagus
    - **B** heart
    - **C** shoulder
    - **D** stomach

Questions 43–45 refer to the following information and graph.

Gauges collect data about the height of the water. The following graph shows the gauge height during high and low tides in the Manasquan River at Point Pleasant, New Jersey, over a period of about a week.

Manasquan River at Point Pleasant, New Jersey

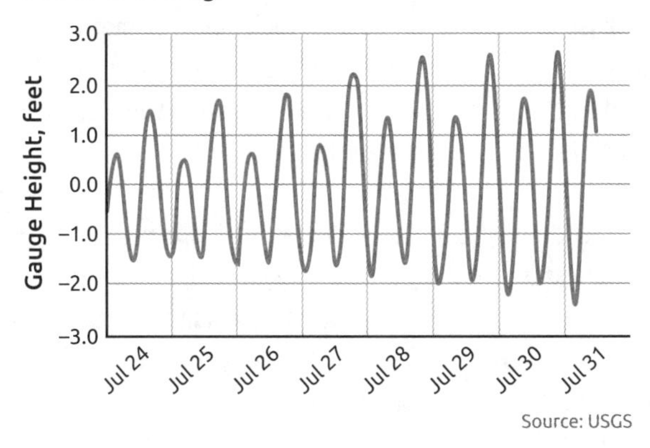

Source: USGS

43. What trend can be determined from the information in the graph?
    - **A** The highest high tide was on July 30.
    - **B** Each day had two high tides and two low tides.
    - **C** The lowest low tide was on the same day as the highest high tide.
    - **D** The mornings had only high tides, while the evenings had only low tides.

44. Which word best describes the information in the graph?
    - **A** cyclical
    - **B** direct
    - **C** inverse
    - **D** linear

45. Which of the following statements is a prediction that can be made based on the information in the graph?
    - **A** July 31 will have a second high tide.
    - **B** July 31 will have two more low tides.
    - **C** The second low tide on July 31 will be at midnight.
    - **D** The highest high tide on July 31 has already occurred.

46. Which of the following organelles is seen in plant cells but not in animal cells?
    - **A** chloroplast
    - **B** cell membrane
    - **C** nucleus
    - **D** mitochondria

47. What is the main way that astronomers classify galaxies?
    - **A** by their number of stars
    - **B** by their age
    - **C** by their shape
    - **D** by their distance from Earth

1. **B.** Both lizards and fish are cold blooded. This means that their internal body temperature is the same as that of their environment.

2. **C.** Only answer C reduces the use of a nonrenewable resource (natural gas). If a restaurant replaces paper cups with plastic cups, it is reducing use of a renewable resource (trees) but increasing use of a nonrenewable resource (plastic made from petroleum).

3. **B.** Subjects in a test group can have differences; a large test group helps account for these differences.

4. **B.** Covalent bonds form between two nonmetals. Lithium and potassium are metals; chlorine and bromine are nonmetals.

5. **C.** Meiosis is the process of cell division by which the gametes, such as eggs and sperm, are produced.

6. **B.** $F = ma$, so $0.5 \times 18 = 9$.

7. **B.** High tide occurs on the side of Earth nearest to the moon (point 3), because the pull of the moon's gravity on the oceans is strongest on that side. Another high tide occurs on the opposite side of Earth (point 1) because the moon's gravity is pulling on Earth harder than it is pulling on the ocean on the far side of Earth, causing the water to pile up on the far side. Low tides occur on the other two sides of Earth, including point 2.

8. **A.** Decomposers are living things, and living things are biotic factors in an ecosystem.

9. **D.** The dependent variable is the variable that changes because of changes in the independent variable.

10. **B.**

11. **C.** An observation is information gathered using the senses.

12. **C.** Star A, the red supergiant, formed from a more massive star. It will eventually explode into a supernova, then cool to form a neutron star or collapse to form a black hole. The red giant formed from a less massive star. It will eventually become a planetary nebula, then a white dwarf, then a black dwarf. One approach to this question would be to sketch out the lifecycle of each type of star, and then select the answer that matches both sketches.

13. **D.** The grasshopper is preyed upon by the toad; if the toads were taken out of the environment, the grasshopper's numbers would increase.

14. **A.** The grass is a producer because it is able to make its own food.

15. **D.** $W = Fd$ (force times distance), so A and B are 0 J (no distance moved), C is 150 J, and D is 200 J.

16. **B.** Temperature decreases with height in the lowest layer of Earth's atmosphere, the troposphere. In the next layer up, the stratosphere, temperature increases with height. In the third layer, the mesosphere, temperature again decreases with height. Since the balloon recorded temperatures decreasing and then increasing, it must have reached the stratosphere. If it had reached the troposphere, temperatures would have begun to decrease again.

17. **B.** An organism's scientific name is made up of its genus and species. The genus comes first and is capitalized.

18. **B.** The question explains that carbon-14 dating can be used only on the remains of living organisms. Only answer B is the remains of a living thing. This question requires you to remember that a trace fossil is evidence of the presence of a living thing, but it does not contain any parts of the living thing.

19. **D.** The law indicates that matter cannot be created or destroyed, so each atom of reactant will also be in the products.

20. **D.** Capillaries are blood vessels that are only one cell thick; this allows for the exchange of materials between the body tissues and the bloodstream.

21. **B.** Wavelength and frequency are inversely related, while frequency and energy are directly related.

22. **B.** Plants rely on bacteria to convert nitrogen into a form they can use. Rain deposits nitrogen into the soil, but not in a form usable by plants. Leaching typically makes nitrogen less available to plants by moving it into deeper layers of soil below where plant roots can reach. Tilling moves soil around but does not convert nitrogen into a usable form.

23. **C.** The liquid mercury and oxygen gas are both different substances from the starting material. In the other three examples, the appearance of the substances change, but the identity of the substances does not change.

24. **D.** A baby is not born with the ability to communicate using sign language, so this behavior is learned.

Answer Key

25. **C.** An inference is a statement that uses outside knowledge.

26. **B.** Multiple trials ensure that no errors were made during an investigation.

27. **D.**

28. **D.** Research is considered reliable if the same experiment can be repeated to produce the same (or similar) results.

29. **C.** Earth's inner core is made of molten (liquid) metal, while the outer core is made of solid metal. The mantle is made of semi-solid rock, but the top layer of the mantle, the asthenosphere, is liquid. The solid plates of the crust float on the liquid asthenosphere. One way to approach this question would be to look for the layers that are not liquid and underline them. Then find the answer choice that has nothing underlined.

30. **D.** A compound has a specific ratio of elements that have chemically reacted and can be represented with a formula, while a mixture is a simple physical combination of materials.

31. **A.** Bases increase the pH of water, while acids decrease it. A pH of 7.0 is considered neutral.

32. **C.** The water is dissolving the potassium hydroxide, so water is the solvent.

33. **C.** A tsunami is a set of large waves caused by an earthquake focused under the ocean.

34. **C.** Potential energy depends on height, so it decreases as the ball falls. Energy is conserved, so the kinetic energy of the ball must increase as potential energy decreases.

35. **B.** A front is named for the type of air mass that is replacing the one that had been in place. In this case, cold polar air is replacing warm tropical air, which makes the boundary between these air masses a cold front.

36. **B.** Acceleration is the change in velocity divided by the change in time. $A = 2.5\ m/s^2$, $B = 3.3\ m/s^2$, $C = 0\ m/s^2$, and $D = 3\ m/s^2$.

37. **D.** Both Earth and the sun have atmospheres. The sun is made of gas and plasma, and it is much hotter than Earth. Because the sun is much more massive than Earth, its gravitational field is much larger.

38. **B.** Phloem vessels, found in plants, are responsible for moving sugar made in the leaves to other parts of the plant, including the roots.

39. **A.**

40. **C.** Metalloids have some of the properties of metals, such as luster and conductivity, and some of the properties of nonmetals, such as brittleness. This is an element and not a compound, because it is made entirely of the same type of atom.

41. **A.** Seasons are created by the tilt of Earth's axis, which causes the Northern and Southern hemispheres to get different amounts of direct sunlight at different times of year. Day and night are caused by Earth's rotation. The phases of the moon are caused by the moon's revolution around Earth. Solar eclipses are caused by the movement of the moon and Earth in relation to the sun.

42. **C.** Voluntary muscles are under conscious control and are attached to the bones of the body. Involuntary muscles can be found in the digestive and circulatory systems.

43. **B.**

44. **A.**

45. **A.** All the days shown on the graph have two high tides and two low tides.

46. **A.** Chloroplasts are part of plant cells but not animal cells because they are necessary for photosynthesis, a process occurring in plant cells but not in animal cells.

47. **C.** The major categories of galaxies are defined by shape. Our galaxy, the Milky Way, is a spiral galaxy. The other types are elliptical galaxies and irregular galaxies.

Answer Key

Check your answers. Review the questions you did not answer correctly. You can use the chart below to locate lessons in this book that will help you learn more about science content and skills. Which lessons do you need to study? Work through the book, paying close attention to the lessons in which you missed the most questions. At the end of the book, you will have a chance to take another test to see how much your score improves.

Question	Where to Look for Help		
	Unit	Lesson	Pages
1	4	5	136
2	2	4	55
3	1	2	23
4	3	2	80
5	4	2	124
6	3	7	101
7	2	6	62
8	4	12	164
9–11	1	1	18–19
12	2	5	57–58
13–14	4	12	164–165
15	3	8	105
16	2	1	43
17	4	8	148–149
18	2	3	49
19	3	3	84
20	4	6	139
21	3	6	98
22	2	4	53–54

Question	Where to Look for Help		
	Unit	Lesson	Pages
23	3	3	84–85
24	4	5	137
25	1	1	19
26–28	1	3	27
29	2	1	43
30–32	3	4	90
33	2	2	46
34	3	5	93
35	2	2	45
36	3	7	101
37	2	6	62
38	4	3	128
39	1	1	18
40	3	1	77
41	2	1	42
42	4	6	139
43–45	1	4	30–32
46	4	1	122
47	2	5	58

Science Inquiry and Skills

Protesters do not trust GMO research results.

Scientific investigations are often featured in the news, especially when the investigators' results contradict earlier studies. How do you know which ones are right? How can you tell if their conclusions are reasonable? Good science skills, especially good science inquiry skills, will help.

KEY WORDS

- analyzing data
- bias
- conclusion
- control
- controlling the variables
- dependent variable
- direct relationship
- error
- evidence
- fair test
- falsifiable
- hypothesis
- independent variable
- inference
- inverse relationship
- investigation design
- limitations
- linear relationship
- nonlinear relationship
- observation
- prediction
- reliable
- scientific method
- testable hypothesis
- trend
- trial
- variable

Investigation Design

KEY WORDS

- conclusion
- control
- dependent variable
- error
- hypothesis
- independent variable
- inference
- investigation design
- observation
- trial
- variable

When a scientist does an experiment, it is often to find out something that is not yet known. There is no way to look up the answer. Scientists also often repeat experiments that were done by other scientists, in order to verify the results. In both cases, it is important to follow a framework when planning and reporting an investigation. The framework helps the researcher make sure the experiment will clearly answer a question, and it helps other researchers repeat that experiment in their own labs. Details about an experiment will change, depending on the topic, but a good **investigation design** will have the same main elements.

Parts of a Well-Designed Investigation

A researcher is studying daphnia in different environments. Daphnia are small aquatic animals that live in fresh water, such as ponds and streams. They are barely visible with the naked eye but can be seen clearly under a microscope. Their bodies are protected by a thin, clear exoskeleton, making their heart and other organs visible. The researcher observes the heart rate of the daphnia before and after adding the same amount of water at different temperatures to their containers. The temperature of the water in the containers before adding other water was 23°C. She recorded the data from her first three **trials** in Table 1.

Change in Heart Rate (bpm)				
	4°C	15°C	23°C	37°C
Trial 1	-30	-18	-6	30
Trial 2	-36	-36	12	33
Trial 3	-12	-6	-3	48
Average	-26	-20	1	37

Table 1: Daphnia heart rate measurements with the addition of different temperatures of water

In her report, the researcher wrote that she predicted the daphnia's heart rate would increase as temperature increased and decrease as temperature decreased. She said that, according to her data, her prediction was correct. "The solubility of oxygen decreases as temperature increases," she wrote. "The daphnia's heart rate increases with increasing temperature to make up for the lowered amount of oxygen in its blood."

The researcher's prediction was her **hypothesis**. A hypothesis can usually be restated as an if/then statement, such as "If I increase the temperature of the water, then the daphnia's heart rate will increase." The **variables** in the experiment were the temperature of the water and the heart rate of the daphnia. The **independent variable** is the variable changed by the researcher. In this case, it was the temperature of the water. The **dependent variable** is the variable that changes because of changes in the independent variable. In this case it was the heart rate of the daphnia. The researcher kept any other variables, such as the amount of water added and the type of organism, constant.

The researcher repeated her procedure enough times that she had three trials at each temperature. Increasing the number of trials increases the accuracy of the data because it reduces the effect of **errors** in measurement. The researcher included a trial in which she did not change the temperature of the water (23°C) she added. This trial was a **control**, to take measurements on a sample without changing the independent variable. The control showed that adding water to the container did not, by itself, change the heart rate of the daphnia. A **conclusion** states whether the hypothesis was correct or incorrect. For a hypothesis to be correct, it must be supported by the data. The researcher concluded that the data supported her hypothesis because the daphnia's heart rate increased with temperature, so her hypothesis was correct.

Observation and Inference

When the researcher counted the heartbeats of the daphnia, she made an **observation**. Observations are information gathered using our senses, such as sight and hearing. When the researcher said that the heartbeats increased because of reduced oxygen in the water, she made an **inference**. Inferences are statements about observations that include additional knowledge. The researcher used her knowledge about the relationship between temperature and the solubility of oxygen in water to make the inference.

Lesson Practice

TEST STRATEGY

Cover the answers to the question. Think of what you would write if the answers weren't there, and then uncover the answers. Choose the one that best matches your own answer.

Complete the activities below to check your understanding of the lesson content.

Skills Practice

Use the following description of an experiment to answer the questions.

A student wanted to see if carbonating water, to make soda water, made it more acidic. He added four different amounts of carbon dioxide to 100 mL of distilled water at 25°C and then measured the pH of the solution. He mixed five solutions for each concentration. The pH of the water with no carbon dioxide was 7.0, and the pH for the highest concentration was 3.0. A pH of 7.0 is neutral; the lower the pH of the solution, the more acidic it is. The student wrote that carbonating water did make it more acidic, because the more carbon dioxide he added, the more the pH of the solution decreased. He finished by saying that the pH decreased because the carbon dioxide reacted with the water to produce carbonic acid.

1. What inference did the student make in his report?
   - **A** that adding carbon dioxide to water lowers the pH
   - **B** that the pH was lowered by the carbonic acid that formed
   - **C** that the lower the pH of a solution, the more acidic it is
   - **D** that carbon dioxide dissolves in water

KEY POINT!

The researcher knows the values of the independent variable before starting the experiment.

2. How many trials did the student run in the experiment?
   - **A** 4
   - **B** 5
   - **C** 10
   - **D** 20

3. What was the student's conclusion?
   - **A** that carbon dioxide forms carbonic acid in water
   - **B** that different solutions of carbonated water have different pH values
   - **C** that more acidic solutions have a lower pH
   - **D** that carbonating water makes it more acidic.

KEY POINT!

The control is a trial in which the independent variable is not changed.

4. Which was the independent variable in the experiment?
   - **A** the amount of carbon dioxide dissolved in the water
   - **B** the pH of the solutions
   - **C** the temperature of the water
   - **D** the pH of pure water with no carbon dioxide

5. What is the student's hypothesis?

____

____

6. Which variables did the student keep constant?

____

____

7. What was the control in this experiment?

____

____

8. Why did the student do so many trials?

____

____

9. What is the difference between an independent variable and a dependent variable?

____

____

10. What is the difference between an observation and an inference?

____

____

See page 37 for answers and help.

Analyzing Research

KEY WORDS

- controlling the variables
- fair test
- falsifiable
- limitations
- scientific method
- testable hypothesis

One challenge of doing good research is posing a good question. A good question has to be relevant and important. It is a researcher's job to understand these elements and formulate a proper question. A good question also needs to be interesting to both the researcher and the scientific community. Conducting experiments and research is a long and tedious process. If the researcher does not find the work exciting, it can negatively influence the process of research. A good question seeks to build upon the previous work of other researchers and to move the understanding of the scientific community forward.

Choosing a Research Question

An eight-year-old boy visits the hospital because of an ear infection. He is experiencing a fever and pain in his ear. He has been taking antibiotics for a few weeks. The antibiotics are needed to control the infection. His mother is concerned about the long-term use of antibiotics and their potential risks. Long-term use of antibiotics can create digestive issues, such as diarrhea. Doctors are asking questions and exploring alternatives. They have to consider the effectiveness of other drugs, such as anti-fungal drugs. Doctors are motivated to form a hypothesis and run an experiment to explore the possibility of finding an alternative.

Doctors have to formulate an appropriate research question. An appropriate research question is a **testable hypothesis**. A testable hypothesis or question can be answered through investigations. The investigations involve experiments, observations, or surveys. A testable hypothesis has variables that can be manipulated and measured. A testable hypothesis relies on evidence instead of personal opinion or belief. A testable hypothesis also needs to be **falsifiable**, meaning that it is possible to reach results that prove the hypothesis to be false.

Using the given information and observations, doctors pose a research question. For example, they might decide to formulate the hypothesis as "Can the anti-fungal drugs treat the ear infection faster than the antibiotics?" This hypothesis is testable, as it can be answered through designing an experiment. It has independent and dependent variables. The independent variable is the type of drug used (anti-fungal versus antibiotics). The dependent variable is the time it takes to treat an ear infection. It is also falsifiable, as the result could show that the anti-fungal drugs do not cure the ear infection faster.

A hypothesis is essentially an educated guess. It involves a question that can be rewritten as "If A happens, then B will happen." This will allow a testable hypothesis to form. In this case, doctors have formulated a hypothesis that can be rewritten as "If anti-fungal drugs are used, then the ear infection will be treated faster."

Choosing the Best Procedure

In experiment design, researchers think about methods that evaluate the hypothesis. These methods have to allow for the variables to be manipulated in order to produce analyzable data. The procedures for the experiment have to be designed to follow the **scientific method**. In the steps of the scientific method, after forming a hypothesis, the researchers use an appropriate procedure to produce results. In this case, the time of treatment for two types of drugs is being tested. Therefore, the procedure needs to include ways to measure this time.

After reviewing previous research, doctors decide to recruit 50 subjects who are willing to participate in the study. They assign an ID number to each participant and use a random number generator to divide the group in half. The first group is assigned to an antibiotics treatment. The second group is assigned to an anti-fungal treatment. It is essential for the subjects to be randomly assigned to each treatment to reduce any risk of bias in the experiment. Both groups begin their treatment at the same time, and their progress is observed and recorded.

It is important for the experiment to be a **fair test**. In a fair test, only one variable is changed, while other conditions are the same. Doctors try to control all other variables by mandating the same diet for the participants. They also monitor the subjects' health history. During recruiting, they select only children between the ages of 6 and 10. This is called **controlling the variables**, during which every condition that is different is accounted for and, if possible, held constant.

Limitations

Every experiment involves many **limitations**. Some of these limitations are about parts of the experiment that cannot be controlled. For example, doctors cannot control every condition of the subjects during the experiment. Some subjects might suffer from other health problems that impact the effectiveness of the drugs. This is why having a larger number of test subjects can help account for differences among the subjects.

Other limitations are about the types of answers that a study can produce. Every study seeks to answer a very specific question; in order to do so, it has to leave out many questions. For example, doctors will not be able to draw any conclusions about the effectiveness of other drugs except for the ones tested. They also will not know how quickly the infection would have cleared up without the help of any drugs. A good researcher is aware of the specific answers he or she is seeking and recognizes the limitations of the experiment.

Lesson Practice

TEST STRATEGY

Underline the key words and phrases in the question and passage. For items with a passage, underline the key words and phrases in the question and then look for them in the passage.

KEY POINT!

A testable hypothesis is narrow enough to be tested and has variables that can be measured.

Complete the activities below to check your understanding of the lesson content.

Skills Practice

Use the following description of an experiment to answer the questions.

Biochar is an organic carbon-rich product that is made when wood burns in a closed container with little air. A student knows that adding other organic material to potting soil will improve it, so she designs an experiment to test the effects of biochar on potting soil. She fills nine identical pots with the same soil mix. Then, she adds 1 gram of biochar to three pots and 5 grams of biochar to another three pots. She plants a bean seed from the same packet of seeds in all nine pots. She places all pots on the same tray and gives each plant the same amount of water every day. She plans to measure the height of each plant after 2 weeks and record the results. It takes 5 weeks for this type of bean to yield fruit.

1. Which statement could be a testable hypothesis for this experiment?
   - **A** Biochar helps bean plants germinate faster if water activates it.
   - **B** Biochar helps bean plants grow larger fruit.
   - **C** Biochar helps bean plants develop deeper roots.
   - **D** Biochar helps bean plants grow taller in the early stages of growth.

2. Each of the following is a constant condition in this experiment except which one?
   - **A** the amount of biochar added to each pot
   - **B** the type of seed planted in each pot
   - **C** the amount of soil added to each pot
   - **D** the size of the pot for each plant.

3. Which would be the best procedure to determine if biochar also makes plants more resistant to pests?
   - **A** observing the growth pattern of the plants if pests attack them
   - **B** introducing one type of pest to the six plants with biochar added to their soil
   - **C** measuring the height of the plants when pests are introduced only to plants with no added biochar
   - **D** introducing one type of pest to all the plants and measuring their height over time

4. Would you describe this study as a fair test? Why or why not?

5. What changes, if any, should the student make to the procedure so the experiment could be used to test a hypothesis about the fruit yield of the bean plant?

See page 37 for answers and help.

KEY POINT!

In a fair test, only one variable is changed while other conditions remain the same.

Evaluating Results

KEY WORDS

- analyzing data
- bias
- conclusion
- evidence
- reliable

A research project is not complete without carefully looking at the gathered data. The data collected from a well-designed investigation is very valuable. This information is used to produce knowledge and to move our scientific understanding forward. Once data is gathered, it is a resource that can be used by scientists for generations to come. This is why it is important to produce reliable data and to properly interpret the data. Likewise, as a researcher using your peer's data, you always need to validate the research, the research process, and the resources used by the researcher.

Analyzing Data

A freshwater conservation institute is running some trials for growing Atlantic salmon in freshwater closed-containment systems. The trials took place on two strains: St. John River and Cascade. The average weight of the strains was measured during different stages of life. The results are shown in the following table.

Strain	St. John River	Cascade
Stock Growout	0.34 kg	0.76 kg
First Early Harvest	2.7 kg	2.6 kg
Last Early Harvest	3.7 kg	2.6 kg
First Premium Harvest	4.2 kg	4.1 kg
Last Premium Harvest	4.7 kg	5.7 kg

The hypothesis was formulated as "Cascade strain produces heavier fish than St. John River." The data reveals the harvest sizes during different times. From this data, the weight of the strains during each harvest can be compared. The table provides **evidence** that needs to be interpreted. Once the data has been interpreted, conclusions can be made based on the interpretations.

Analyzing data is the process of inspecting and interpreting data to discover useful information for drawing conclusions. Here, the data reveal that the St. John River strain is heavier during the Stock Growout stage. By the First Early Harvest, the two strains have almost the same weight. During the Last Early Harvest, the weight of St. John River increases, while Cascade does not grow. During the First Premium Harvest, the weights match each other once again. However, for the final harvest, Cascade shows a significant increase over St. John River.

Drawing Conclusions

One of the most important stages of scientific research is using the data gathered to draw conclusions. A set of data provides a large body of information. A good researcher keeps the hypothesis in mind while suggesting conclusions for the research. A **conclusion** is the answer to a testable question that is supported by collected evidence and data. The evidence gathered during one experiment could support or cast doubt on the hypothesis. It is rarely the case that the evidence could prove or disprove the hypothesis. This is due to limitations of an experiment and the requirements that make it specific enough to be testable. Therefore, in interpreting data, it is important to use the correct language about the evidence being in agreement or disagreement with the hypothesis. Absolute language, such as "the evidence proves that the hypothesis is correct" is best avoided.

According to the analyzed data, St. John River and Cascade salmon gain weight differently over time. While St. John River grows heavier at least at one point, this experiment shows that the weight of Cascade is indeed more than St. John River during the Last Premium Harvest. It is important to recall the hypothesis: "Cascade strain produces heavier fish than St. John River." The evidence suggests that the weight of Cascade is not always heavier than St. John River. In fact, out of five stages of harvest, only during two stages is Cascade heavier. Therefore, the hypothesis is partially supported by the evidence.

Bias

Researchers have to watch for potential bias in design procedure and drawing conclusions. **Bias** is an approach that is motivated by one's opinion or belief instead of by evidence. Researchers have techniques, however, to reduce the amount of bias involved in design. One of these techniques is random selection of subjects of study. This might happen using a random number generator or another method that could eliminate the chance of biased selection.

For example, a biased researcher in this experiment might be looking for the heavier fish to weigh in order to increase the chances of supporting evidence. A better method could be using ID numbers for the fish and using a random number generator to select a few IDs. Then, the researcher could take the average weight of the randomly selected fish.

Biased interpretations sometimes take place while drawing conclusions. For example, a biased researcher might analyze the data and imply that the evidence proves the hypothesis. The researcher would be ignoring the fact that harvest at three out of five stages indicated lighter weight for Cascade salmon. To avoid biased conclusions, a good researcher considers parts of the hypothesis that are being supported or doubted according to the evidence. Remember, even when all the evidence is in agreement with a hypothesis, one experiment is rarely enough to prove a hypothesis.

Reliability

Research is considered **reliable** if the same process can be repeated by other scientists under the same conditions to produce the same results. Reliable research reaffirms the findings. It ensures that the scientific community will accept the results and that the research results could be used as evidence in future research. One way to increase reliability is to employ multiple trials.

Reliable research is built upon other reliable resources. As a researcher, it is essential to look for credible resources for gathering data. For example, resources such as blogs or journals that are based on personal opinion or belief are not considered reliable. On the other hand, the results of an experiment that

Evaluating Results

follows the scientific method is a reliable resource. Peer-reviewed scientific journals include studies that are reliable. This means that the studies not only meet the requirements of conducting reliable scientific research, but they have also been evaluated by at least one person with academic competency similar to the researcher. Peer-reviewed scientific work guarantees the production of quality scientific work that is considered reliable and credible.

Lesson Practice

Complete the activities below to check your understanding of the lesson content.

Skills Practice

Use the following description of an experiment to answer the questions.

A teacher is testing a hypothesis that eating a small snack before taking a test can improve test scores. In order to test the hypothesis, the teacher recruited five students to participate in the study. First, the teacher recorded the test results of the five students having had no snack. Prior to another test, the teacher provided them with a snack. The teacher recorded the results in a table:

Student #	Test Score When There Was No Snack	Test Score When Students Snacked
1	84	82
2	90	95
3	83	85
4	78	75
5	92	96

1. Which analysis is correct based on the data in the table?
   A The number of students whose test score increased is higher than that of those whose score decreased.
   B The number of students whose test score decreased is higher than that of those whose score increased.
   C The total increase in test score is less than the total decrease before and after snacking.
   D The total decrease in test score is equal to the total decrease before and after snacking.

TEST STRATEGY

Treat each answer option as a true/false question. Jot down why each is true or false.

2. Is the evidence in support of the hypothesis? Why or why not?
   - **A** No, because the score of some of the students dropped after snacking.
   - **B** No, because the average decrease in score is higher than the average increase in score.
   - **C** Yes, because the scores of all students increased after snacking.
   - **D** Yes, because the average increase in score is higher than the average decrease.

3. How does the evidence gathered in the experiment relate to the statement "Snacking improves test scores"?
   - **A** The data disproves the statement.
   - **B** The data supports the statement but does not prove it.
   - **C** The data proves that the statement is correct.
   - **D** The data casts doubt on the statement but does not disprove it.

4. What are some possible sources of bias in the procedure the teacher chose?

____

____

____

____

5. How can the teacher increase the reliability of her research?

____

____

____

____

See page 37 for answers and help.

KEY POINT!

Bias is an approach that is motivated by one's opinion or belief instead of by evidence. Using strong and definite language in data analysis is often a sign of bias.

KEY POINT!

An experiment is reliable if the same process can be repeated by other scientists under the same conditions to produce the same results.

Interpreting Data

KEY WORDS

- direct relationship
- inverse relationship
- linear relationship
- nonlinear relationship
- prediction
- trend

When scientists conduct investigations, they gather a great deal of data. Researchers need a way to better understand this information. Using graphs, tables, and charts enables scientists to see trends, determine relationships, and make predictions.

Tables

Tables are a useful way for scientists to organize data. Take, for example, the times and heights of high and low tides over the course of one week. Having this information displayed in a list would be cumbersome and difficult to read. Organizing the data into a table, however, allows the information to be analyzed; scientists can use the information to determine **trends** and relationships.

	High		Low	
	AM	PM	AM	PM
1. Mon	3:00	3:29	9:20	9:42
2. Tue	4:01	4:27	10:23	10:37
3. Wed	4:58	5:22	11:20	11:29
4. Thu	5:51	6:13	11:55	12:13
5. Fri	6:40	7:01	12:18	1:03
6. Sat	7:25	7:47	1:05	1:50
7. Sun	8:09	8:32	1:50	2:34

Using the data in the tide table, you can determine that tides are cyclical in nature, with two high tides and two low tides each day. You can see that the difference between the two high tides and the two low tides is approximately 12 hours.

Charts

Charts are another useful way to organize and lay out data and information. A commonly used type of chart is called a pie chart; each section of the "pie" represents a different piece of data. This pie chart demonstrates water quality ratings at Volunteer Creek.

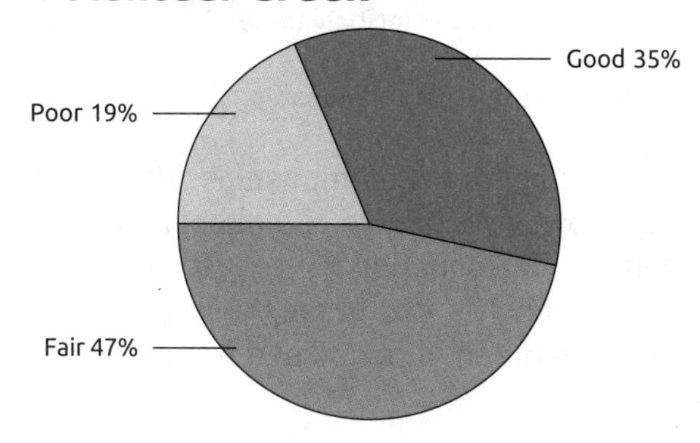

Graphs

Graphs are a method used to determine trends and relationships in data by allowing scientists to compare information in a visual way. There are several types of graphs, all of which have two axes, an *x*-axis and a *y*-axis. One type of graph, a line graph, is used here to show the relationship between time period and frequency for electromagnetic waves.

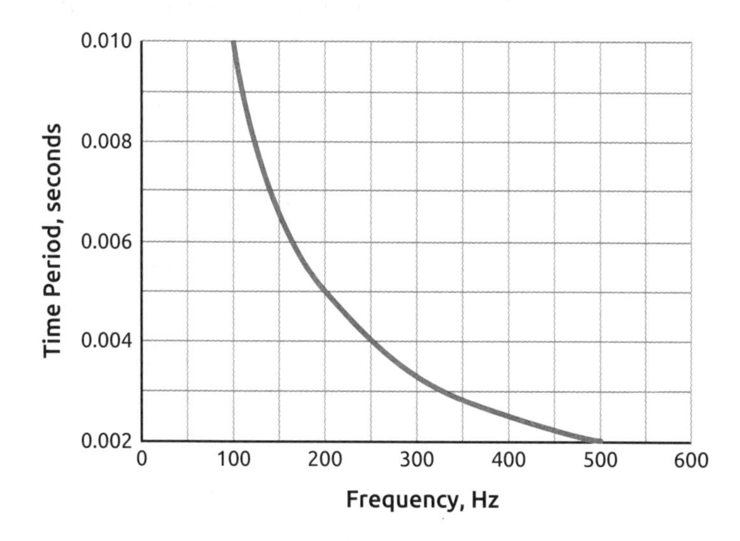

On this line graph, the frequency is on the *x*-axis, and the time period is on the *y*-axis.

Because the line is not a straight line, this graph demonstrates a **nonlinear relationship**. Nonlinear relationships indicate that the rate of change (the slope of the line) is increasing or decreasing. A graph that shows the relationship between two variables as a straight line demonstrates a **linear relationship**. The rate of change stays constant in a linear relationship.

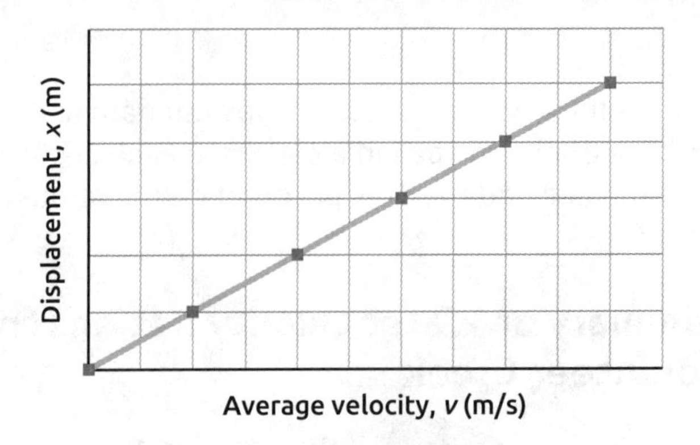

This graph shows a linear relationship.

Linear relationships are also called **direct relationships**. The graph of displacement and velocity shows an example of a direct relationship. These graphs can be represented with the equation $y = mx + b$. **Inverse relationships** can be represented by the equation $y = k/x$, where the x values increase faster than the y values decrease. The graph of period and frequency shows an example of an inverse relationship.

Making Predictions

An important part of a scientific investigation is making a **prediction** based on data gathered. Studying tables, charts, and graphs allows researchers to be better able to see trends, which in turn allows them to make predictions. For example, with the tide chart, we can predict that the first high tide on the Monday of the next week will be between 8:45 a.m. and 9:00 a.m. Likewise, with the frequency graph, we can predict that a wave with a frequency of 50 Hz would have a period of about 0.02 seconds.

Complete the activities below to check your understanding of the lesson content.

Skills Practice

Answer the questions based on the content covered in the lesson.

1. Dr. Cohen collected data during an investigation and organized it in the following table:

x-values	y-values
1	2.00
3	0.67
4	0.50
7	0.28
9	0.22

Which of the following terms BEST describes the relationship between the variables?

A comparative
B direct
C inverse
D linear

Use the following graph to answer questions 2–4.

Effect of Temperature on Volume of a Gas

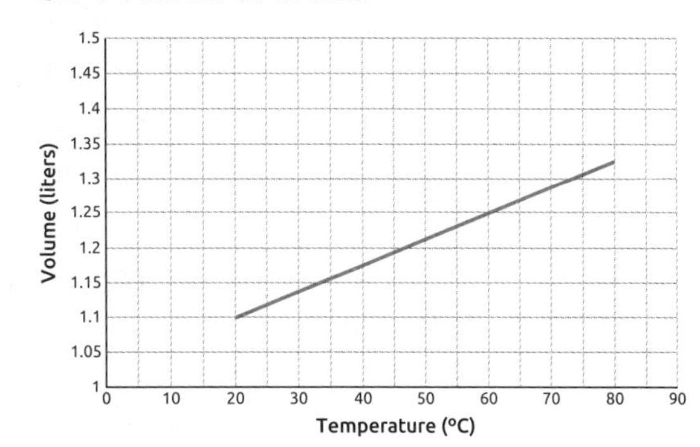

2. What type of relationship does this graph show?

A direct but not linear
B inverse but not linear
C direct and linear
D inverse and linear

3. Which variable is on the *x*-axis?

____

4. What do you predict will happen to the volume of the gas at 90°C?

____

____

____

See page 38 for answers and help.

KEY POINT!

Remember that in a direct relationship, both variables increase at the same rate.

TEST STRATEGY

Using a diagram can help you visualize the question. Draw a graph with the *x* and *y* values to help you find the answer. What is the relationship between the variables?

Answer the questions based on the content covered in this unit.

Use the passage and graph below to answer questions 1–4.

Scientists predict that global temperatures will increase in the next century and that the increase will be greatest in the boreal and subarctic regions. They also predict that the climate in those areas will become drier. Peatlands, such as bogs and fens, make up 15 percent of these areas, but they contain over one-third of the world's soil carbon (mostly in the form of decomposed plant material). Some of this material becomes dissolved in rainwater and is called dissolved organic carbon (DOC). As water drains from the peatlands into nearby rivers and lakes, a small, constant amount of DOC goes with it. At low concentrations, DOC has been found to be beneficial to the ecosystems of these waterways, as it filters out UV light, controlling algae growth, and removes from the water toxins that can harm fish.

A team of researchers at the University of Minnesota found that raising the temperature and lowering the amount of water available to bogs and fens decreased the amount of DOC being released. They used heat lamps and drains to simulate the effects of global temperature increase on one bog. This decreased the amount of DOC being released. They then measured the amount of carbon dioxide and other carbon-containing gases being released from the bog. They compared the number of grams of carbon gases released for each square meter of bog per year with the decrease in the grams of DOC per square meter per year.

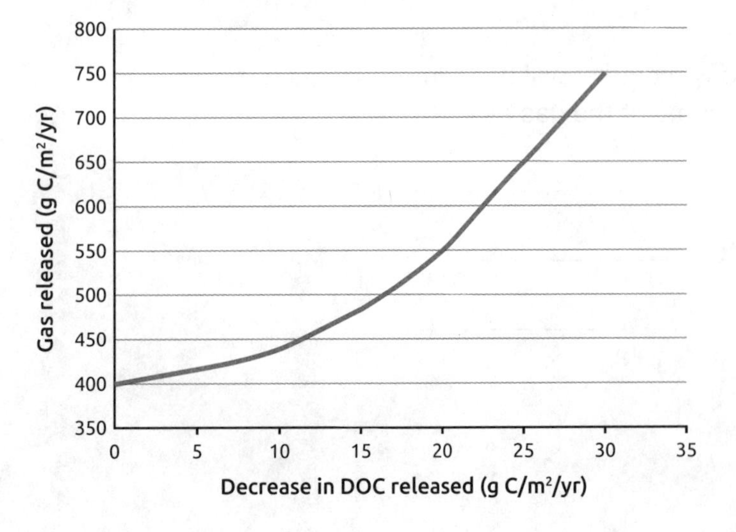

1. Which statement is a testable hypothesis for the procedure described in the passage?
   - **A** Raising the temperature and lowering the amount of water available to bogs decreases the amount of DOC being released.
   - **B** The effects of global warming on bogs can be simulated using heat lamps and drains.
   - **C** As the amount of DOC decreases, more carbon dioxide and other carbon-containing gases will be released from bogs.
   - **D** DOC in small amounts is beneficial to the ecosystem because it controls algae growth.

2. One of the scientists stated in the report that the increase in carbon-containing gases is due to higher temperatures speeding up the decomposition of the DOC retained. Which BEST describes this statement?
   - **A** It is an inference, because the study did not measure temperature.
   - **B** It is an observation, because the scientist knew that heat speeds decomposition.
   - **C** It is an inference, because the study did not measure decomposition rates.
   - **D** It is an observation, because the scientist knew that decomposition produces gases.

3. Which of these could be a source of error in the experiment?
   - **A** measuring the amount of DOC released by the bog
   - **B** determining the type of bog being studied
   - **C** identifying the composition of the gases released
   - **D** choosing the correct heat lamps to use

4. Which of the following correctly describes the statement "Global warming will cause peatlands to release more carbon dioxide and other greenhouse gases"?
   - **A** The data prove the statement is correct.
   - **B** The data support the statement but do not prove it is correct.
   - **C** The data do not support the statement and do not prove it is correct.
   - **D** The data prove the statement is not correct.

Use the graphs below to answer questions 5–9.

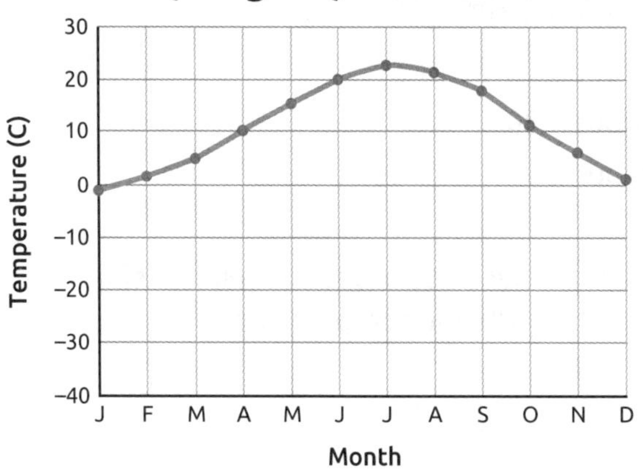

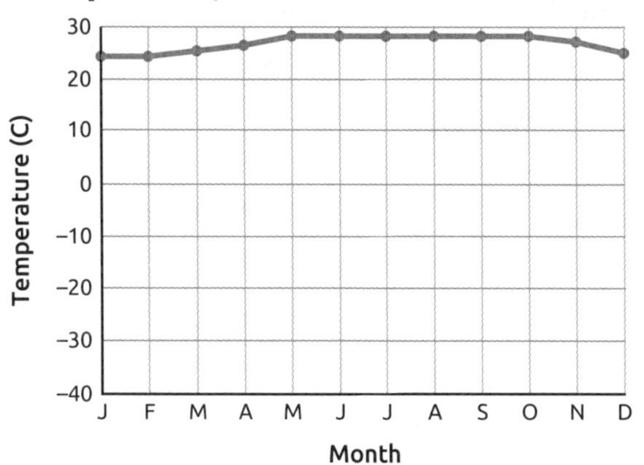

5. Which statement describes how the yearly temperatures in Campa Pita, Belize, compare to the yearly temperatures in Staunton, Virginia?
   - **A** In Campa Pita, temperature varies according to month of the year. In Staunton, there is no relationship between temperature and month of the year.
   - **B** In Campa Pita, temperature is directly related to month of the year. In Staunton, temperature has an indirect relationship with month of the year.
   - **C** In Campa Pita, temperature stays fairly constant over the course of a year. In Staunton, temperature is cyclical, with higher temperatures in the summer and lower temperatures in the winter.
   - **D** In Campa Pita, temperature ranges greatly over the course of a year, with temperatures increasing and decreasing. In Staunton, temperature stays within a range of several degrees over the course of a year.

6. Leanne is traveling to Staunton, Virginia, in August. What can she expect the average temperature to be during her visit?
   - **A** around 0°C
   - **B** around 10°C
   - **C** around 20°C
   - **D** around 30°C

7. Which term best describes the relationship between month and temperature for both locations?
   - **A** direct
   - **B** inverse
   - **C** linear
   - **D** nonlinear

8. Which month in Staunton, Virginia, has the lowest average temperature?
   - **A** November
   - **B** December
   - **C** January
   - **D** February

9. Which two months in Staunton, Virginia, have about the same average temperature?
   - **A** December and April
   - **B** October and April
   - **C** July and September
   - **D** February and September

See page 38 for answers.

Lesson 1

1. B. The student used outside knowledge that carbon dioxide reacts with water to make carbonic acid—this was not something he could observe.

2. B. The student did five trials for each concentration. Don't confuse the number of times he changed the independent variable with the number of trials—trials are "repeats."

3. D. The student's conclusion states whether his hypothesis was right, so you must first determine the hypothesis. Look for what the student was trying to learn by doing the experiment.

4. A. The independent variable is what the student changed; he knew what each value would be before the experiment because he chose the amounts of carbon dioxide he would dissolve.

5. Sample answer: Carbonating water makes it more acidic. The if/then statement would be: "If I add carbon dioxide to water, then the pH will decrease."

6. Sample answer: The student kept the starting pH of the water, the type of water, the temperature, and the amount of water used constant. You can find the variables that were kept constant in the description of the setup. For example, the setup mentioned the temperature of the water.

7. Sample answer: The control was the water with no carbon dioxide mixed in.

8. Sample answer: The student did five trials because every measurement he makes, such as the amount of water, amount of carbon dioxide, or the pH measurement, can be slightly inaccurate. Increasing the number of trials reduces the effects of these errors.

9. Sample answer: The independent variable is the one changed by the researcher. The researcher knows ahead of time what these values will be, as they are planned as part of the design. The dependent variable is what the researcher measures after changing the independent variable. The dependent variable values are unknown until the researcher completes the experiment.

10. Sample answer: An observation is a report of what the researcher saw, or what was recorded by instruments used in the experiment. It requires no knowledge to report. An inference uses other information, such as facts relating to the area of study. It goes beyond what information was gathered in the experiment.

Lesson 2

1. D. The student is testing only a specific plant during a specific period of time. The experiment does not measure the time of germination or the development of the root. Since the experiment runs for only 2 weeks, fruit production is also not measured.
2. A. This is the only variable that is changed among the options.
3. D. Since the amount of biochar is the independent variable and height is the dependent variable, all others must be constant, so all plants must receive the same pest.

4. Sample answer: Yes, it is a fair test because only the amount of biochar is different while other conditions are the same.

5. Sample answer: The plants produce fruit in 5 weeks. This setup can be used if the experiment runs longer and the results are recorded. The number of fruit each plant yields, size of fruit, time before fruit appears, and time before fruit is fully grown all could be measured.

Lesson 3

1. A. Since the test score increases for three students and decreases for two students, the first statement is correct. Statement B is false as it is the opposite of this finding. Statements C and D are comparing the total increase or decrease. Total increase for three students is 11 points. Total decrease for two students is 5. This makes statements C and D false.

2. D. This analysis of the data is in support of the hypothesis. The average increase in score is 3.6 because the total increase is 11 divided by 3 students. The average decrease is 2.5, since the total decrease is 5 divided by 2 students.

3. B. The data shows an average improvement in test scores for 3 out of 5 students. This evidence is in support of the statement that "Snacking improves test scores," but the evidence from one experiment is not sufficient to prove the statement.

4. Sample answer: The students were not selected randomly. The test difficulty varied between the test with no snack and the test when students snacked.

5. Sample answer: by increasing the number of trials or including more students in the experiment

Lesson 4

1. C. The *y* values should decrease as the *x* values increase; you should also notice that as the *x* values decrease, the change in *y* values becomes smaller.

2. C.

3. temperature in degrees Celsius

4. Sample answer: The volume will increase to about 1.36.

Unit Test

1. C.
2. C.
3. A.
4. B.
5. C.
6. C.
7. D.
8. C.
9. B.

Unit Glossary

- **analyzing data** – the process of inspecting and interpreting data to discover useful information about drawing conclusions
- **bias** – anything that sways an experiment's results in a way that makes them inaccurate
- **conclusion** – the solution or answer to the hypothesis
- **control** – factors that are kept the same or constant during an experiment
- **controlling the variables** – the process during an experiment in which every condition that is different is accounted for and held constant
- **dependent variable** – factors that are changed in response to the independent variable during an experiment
- **direct relationship** – a relationship between variables in which an increase or decrease in one variable causes the other variable to do the same
- **error** – a mistake
- **evidence** – information or results that support or counter a hypothesis
- **fair test** – an experiment in which only one variable is changed
- **falsifiable** – it is possible to reach results that prove the hypothesis to be false
- **hypothesis** – a prediction or an educated guess that answers a question.
- **independent variable** – factors that are changed by a scientist during an experiment
- **inference** – statements and observations that include additional knowledge
- **inverse relationship** – a relationship between variables in which an increase or decrease in one variable causes the other to do the opposite
- **investigation design** – a procedure designed to answer a scientific question
- **limitations** – parts of the experiment that can be controlled
- **linear relationship** – a relationship between variables that indicates a constant rate of change
- **nonlinear relationship** – a relationship between variables that indicates the rate of change is either increasing or decreasing
- **observation** – information gathered using the senses
- **prediction** – a specific belief about the hypothesis
- **reliable** – what an experiment is considered if the same process can be repeated by other scientists to produce the same results
- **scientific method** – a series of steps used to gather and test information
- **testable hypothesis** – a hypothesis that can be answered though investigation
- **trend** – a pattern/a prevailing tendency
- **trial** – one of a number of repetitions of an experiment
- **variable** – a factor that changes in an experiment

UNIT 1

Study More!

Consider exploring these concepts, which were not introduced in the unit:

- Double-blind, placebo-controlled investigations
- Deductive and inductive inferences
- Statistical controls
- Types of bias (design, sampling, procedural, personal)
- Graphing independent and dependent variables
- Three-variable graphing techniques

UNIT 2

Earth and Space Science

Earth and space scientists study how land, water, and the atmosphere are connected, as well as all the objects in our solar system and beyond. They examine phenomena that are easy to see, such as the cycle of night and day or a sudden thunderstorm. They also explore nearly imperceptible changes, such as the slow building of a mountain, the shifting of continents, or the life cycle of a star.

Earth and Space Science

UNIT 2

KEY WORDS

- air mass
- asthenosphere
- astronomy
- atmosphere
- axis
- big bang model
- black dwarf
- black hole
- carbon cycle
- closed universe theory
- conservation
- continental drift
- core
- crust
- current
- deposition
- earthquake
- elliptical galaxy
- erosion
- exosphere
- flat universe theory
- focus
- fossils
- front
- globular clusters
- gravity
- greenhouse effect
- heliocentric
- humidity
- hurricane
- hydrosphere
- igneous rocks
- irregular galaxy
- lithosphere
- lunar eclipse
- magma
- main-sequence star
- mantle
- mesosphere
- metamorphic rocks
- Milky Way
- natural resources
- nebula(e)
- neutron star
- nitrogen cycle
- nonrenewable resources
- open universe theory
- ozone layer
- Pangaea
- penumbra
- phase
- planetary nebula
- plate tectonics
- pollution
- red giant
- red supergiant
- renewable resources
- revolution
- Richter scale
- rotating
- seasons
- sedimentary rocks
- seismic waves
- solar eclipse
- solar system
- spiral galaxy
- stratosphere
- sunspots
- supernova
- thermosphere
- tide
- troposphere
- tsunami
- umbra
- weathering
- white dwarf

Earth

KEY WORDS

- asthenosphere
- atmosphere
- axis
- core
- crust
- exosphere
- hydrosphere
- lithosphere
- magma
- mantle
- mesosphere
- revolution
- rotating
- seasons
- stratosphere
- thermosphere
- troposphere

Although it seems like a strange comparison, the structure of Earth is much like that of a hard-boiled egg. If one were to divide an egg into two halves, it would be apparent that the egg was composed of several layers, just like Earth.

Earth in Space

It takes Earth approximately 365 days (one year) to complete one **revolution** around the sun. Because Earth is tilted on its **axis**, different parts of the planet are tilted toward and away from the sun as Earth completes its journey around the sun. This phenomenon is responsible for the **seasons**. For example, in the winter, the Northern Hemisphere is tilted away from the sun; in the summer, the Northern Hemisphere is tilted toward the sun and is at its greatest distance from the sun. The areas tilted away from the sun receive less direct sunlight, and the areas tilted toward the sun receive more direct sunlight. The equator receives equal amounts of sunlight throughout the year.

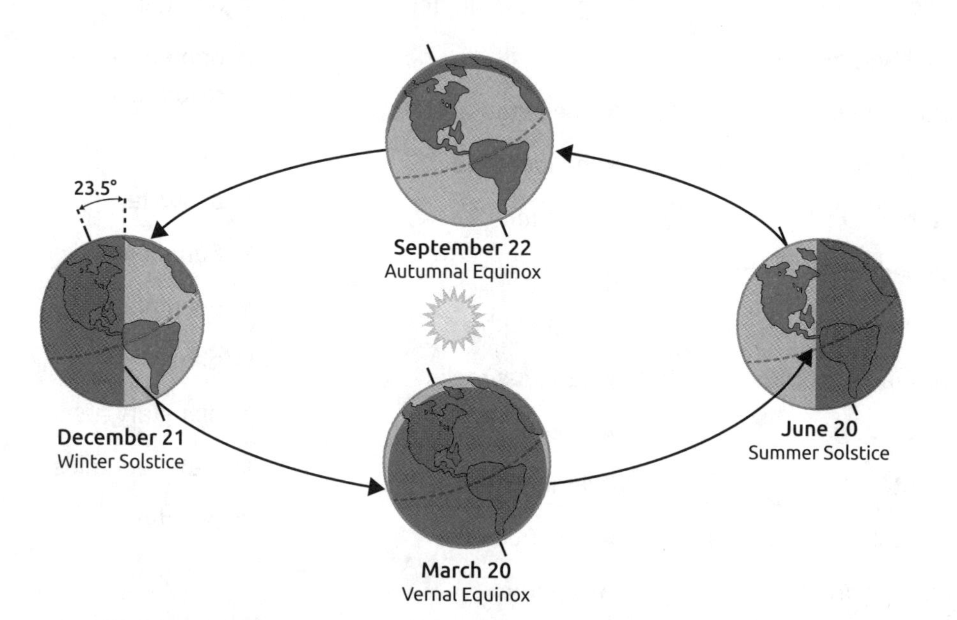

Earth's revolution and the tilt of its axis at 23.5° cause the seasons. During the equinoxes, Earth experiences an equal number of day hours and night hours. During the summer solstice, the Northern Hemisphere experiences its longest day of the year, while during the winter solstice, it sees the lowest number of daylight hours. The dates in this illustration are for 2016, based on Coordinated Universal Time.

At the same time Earth is traveling around the sun, it is also **rotating** on its axis, making one full rotation in approximately 24 hours, or one day. To an observer on Earth, as the planet rotates, the sun and moon appear to move across the sky from east to west.

Earth's Interior

Earth, with a diameter of nearly 8,000 miles, is made up of three distinct layers: the crust, the mantle, and the core. The rigid **crust** is the outermost layer and contains all of Earth's landforms and life. It is also the thinnest layer, with most of it measuring around 3 miles in depth. It can range, however, to around 18 miles beneath oceans and as much as 62 miles beneath large mountain ranges.

Beneath the crust is the **mantle**, which consists of rock in a semi-solid state, called **magma**. Because temperature and pressure increase as depth increases, the mantle is denser and hotter than the crust. It is also thicker, at about 1,800 miles. The cooler, more solid part of the mantle and the crust make up the lithosphere. The **lithosphere** is a rigid layer of rock that is broken up into plates that hold the continents and oceans. These plates are thought to "float" on top of a liquid portion of the mantle called the **asthenosphere**, giving rise to the term "plate tectonics."

At Earth's center, the **core** is made up mostly of the metal iron. It is divided into two parts: the solid, 1,370-mile-thick outer core and the liquid, 780-mile-thick inner core. Earth's rotation causes the liquid inner core to rotate as well, which creates Earth's magnetic field.

The Atmosphere and Hydrosphere

Like the interior of Earth, the **atmosphere** is also made up of distinct layers. The layers change in both composition and characteristics from one to the other. The following table shows the layers of the atmosphere.

Layer Name	Height Above Earth's Surface	Important Facts
Exosphere	400+ km (248.5+ mi)	Molecules escape into space; satellites orbit Earth here.
Thermosphere	320 km (195.6 mi)	Temperatures increase with height; thin layer of gases; contains the ionosphere.
Mesosphere	80 km (49.7 mi)	Gases become denser; slows down meteors; temperatures decrease with height.
Stratosphere	50 km (21.1 mi)	Holds 19% of total gases; temperatures increase with height; bottom layer can be marked with the tops of cumulonimbus clouds.
Troposphere	12 km (7.5 mi)	Density of gases decreases with height; temperatures decrease with height; all weather occurs here.

The **hydrosphere** covers approximately 70% of Earth's surface. It contains all the water on Earth, including oceans, lakes, rivers, ponds, and glaciers.

Lesson Practice

TEST STRATEGY

Rewrite the question in your own words to make sure you understand it.

KEY POINT!

Earth's rotation causes day and night. Earth's revolution and tilt on its axis cause the seasons.

Complete the activities below to check your understanding of the lesson content.

Skills Practice

Answer the questions based on the content covered in the lesson.

1. When the Northern Hemisphere is tilted toward the sun, it is experiencing the season called

   ____

2. Explain how Earth's magnetic field is created.

   ____

   ____

   ____

   ____

3. The layer of the atmosphere in which satellites orbit Earth is the

   ____

4. In which layer of the atmosphere does all weather occur?
   A exosphere
   B mesosphere
   C stratosphere
   D troposphere

5. What makes up the lithosphere?
   A the upper and lower mantle
   B the crust and upper mantle
   C the inner and outer core
   D the crust and outer core

6. Which of Earth's layers contains the oceans and continents?
   A crust
   B mantle
   C inner core
   D outer core

7. If Earth were to stop rotating, what would happen?
   A Earth would have no seasons.
   B Earth would no longer be tilted.
   C One side of Earth would always have daytime.
   D The equator would experience the most direct sunlight.

8. Which of the following factors are responsible for Earth's seasons?
   A revolution alone
   B rotation alone
   C rotation and tilt of axis
   D revolution and tilt of axis

See page 70 for answers and help.

It is late afternoon on a hot, humid summer day. Dark, threatening clouds begin to approach, and rolls of thunder can be heard in the distance. Before long, the sky flashes white with lightning, and raindrops pelt the ground—lightly at first, and then in a downpour. A thunderstorm like this is an example of our ever-changing Earth and atmosphere.

Currents

Different parts of Earth receive different amounts of the sun's energy, causing uneven heating. This results in wind and ocean **currents**. Along the west coasts of the continents, cold currents flow toward the equator, while on the east coasts of the continents, warm currents flow from the equator toward the poles. In the United States, the cold current is known as the California Current, and the warm current is called the Gulf Stream, one of the strongest currents in the world. Currents have a great impact on long-term weather. For example, winters in the United Kingdom are warmer than those of other locations at the same latitude because of the warm water carried by the Gulf Stream.

Air Masses

Air masses, or large bodies of air, contribute to much of the weather we experience each day. Air masses are classified by their characteristics, each mass generally uniform in temperature and **humidity** (the amount of moisture in the air). They are named according to the region in which they formed.

Type of Air Mass	Characteristics
Polar (arctic)	Cold, dry
Continental	Dry
Maritime	Moist
Tropical	Warm, moist

Sometimes, as air masses move, they pick up characteristics of other areas. For example, if a polar air mass moves over the ocean, it can pick up moisture and become a maritime polar air mass.

The boundary where two air masses meet is called a **front**. Fronts are named for the type of air mass that is replacing the one that had been in place. For example, if a cold air mass is moving in behind a warm air mass, the leading edge of the cold air mass is called a cold front. Cold fronts are often associated with a narrow band of showers and thunderstorms, while warm fronts are usually preceded by a few hours of light to moderate rain. If the air masses are not moving, the front is called a stationary front.

KEY WORDS

- air mass
- current
- earthquake
- focus
- front
- humidity
- hurricane
- Richter scale
- seismic waves
- tsunami

Weather and Natural Disasters

Natural Disasters: Extreme Weather

Although thunderstorms occur all over the United States, they are most common in southeastern states. This is because temperature and moisture are the key ingredients in a thunderstorm. Evaporation from the warm air currents on the East Coast puts more moisture into the atmosphere. Warm, moist air near the surface and cold, dry air aloft cause the air to become unstable. As the warm air rises and cools, condensation occurs and forms the cumulonimbus clouds of thunderstorms. Tornadoes can form from thunderstorms known as supercells.

In areas near the equator, conditions are generally warm and humid all year. It is in these areas that tropical cyclones develop. Known as **hurricanes** in the United States, tropical cyclones are large low-pressure systems with organized circulation. They are characterized by high, sustained wind speeds (at least 74 miles per hour) and large amounts of precipitation. These storms can cause devastating floods where they strike.

Natural Disasters: The Shifting Earth

Earthquakes and most volcanic eruptions occur at fault zones. One of the most famous faults in the United States is the San Andreas. Here, the Pacific plate is moving northward relative to the North American plate. This movement can occur either in a slow, steady motion or in sudden bursts. The slow-moving areas experience small to moderate earthquakes. In between these areas, however, are regions of infrequent earthquake activity. These regions build up tremendous amounts of energy for decades or centuries that are released in a sudden burst, resulting in a devastating earthquake. This area of release inside Earth is called the **focus**. The energy released travels through the Earth's interior by **seismic waves**. The magnitude of an earthquake is measured using a method called the **Richter scale**, which ranks earthquakes on a scale of 1 to 10. The Richter scale is logarithmic, meaning that each consecutive ranking is 10 times stronger than the previous one.

An earthquake's focus that is underwater near a marine or coastal area can trigger a **tsunami**, or a set of ocean waves. Some earthquakes indirectly generate tsunamis by triggering underwater landslides. The landslides then trigger tsunamis. The waves generated by tsunamis can be meters high. For example, in 1998 three waves more than 7 meters high struck Papua New Guinea, wiping out three coastal villages within 10 minutes of a 7.0 earthquake.

Complete the activities below to check your understanding of the lesson content.

Skills Practice

Answer the questions based on the content covered in the lesson.

1. A front that precedes a colder air mass is called a ____________________ front.

2. The warm current along the East Coast of the United States is called the ____________________.

3. On March 27, 1964, a magnitude 9.2 earthquake struck southern Alaska. Its effects were felt over 311 miles (500 km) away from the focus. Explain how the earthquake was felt so far away from the focus.

   ____

   ____

   ____

   ____

4. Which two events can generate a tsunami?

   ____

   ____

5. Clarkstown is experiencing several hours of light rainfall. Which type of front is most likely moving through the area?

   A warm front
   B cold front
   C stationary front
   D occluded front

6. Why is the West Coast of the United States more likely than the East Coast to experience a tsunami?

   A The currents on the East Coast are stronger.
   B The air on the West Coast is cooler and drier.
   C The West Coast is closer to a plate boundary.
   D The ocean on the East Coast is deeper and saltier.

7. How do large earthquakes occur?

   A as plates move slowly and gradually past each other
   B as areas that have built up energy for many years release energy
   C as currents move quickly past a fault zone in a coastal area
   D as underwater landslides push water up toward the surface

KEY POINT!

Cold fronts bring in cold air masses, while warm fronts bring in warm air masses.

Lesson Practice

KEY POINT!

Heat and moisture are key ingredients for a thunderstorm.

8. How does condensation contribute to the formation of thunderstorms?
   - **A** Condensation of warm, moist air puts moisture into the atmosphere.
   - **B** Condensation of warm, moist air allows clouds to move higher up into the atmosphere.
   - **C** Condensation of cool, dry air forces cumulonimbus clouds closer to Earth's surface.
   - **D** Condensation of cool, dry air replaces warm air at Earth's surface and forms surface winds.

9. Why do the majority of thunderstorms in the United States occur in the Southeast?
   - **A** It is warm and humid.
   - **B** It is near a fault line.
   - **C** It has the most cumulonimbus clouds.
   - **D** It has many areas of cyclonic activity.

10. As hurricanes travel northward, they tend to become weaker. Which statement best explains why this occurs?
    - **A** Ocean temperatures are cooler at more northern latitudes, and hurricanes derive their energy from warm water.
    - **B** Currents are weaker at more northern latitudes, and hurricanes use the force of currents to move over distances.
    - **C** The Coriolis effect is less noticeable at more northern latitudes, and hurricanes sustain their cyclonic motion from the Coriolis effect.
    - **D** Air masses are stronger at more northern latitudes, and hurricanes are pushed away by the stronger energy of the air masses.

See page 70 for answers and help.

Humans have always tried to understand Earth and how it works. From organic matter and living organisms to mountains and rocks, scientists search for answers about the past, present, and future state of the elements of Earth. They are also interested in understanding why rocks break and move away and why soil washes off a slope, so that they can slow down the process when needed. These questions have led scientists to look for evidence all over Earth to create theories. Today, scientists know that Earth's surface is extremely dynamic.

KEY WORDS

- continental drift
- deposition
- erosion
- fossils
- igneous rocks
- metamorphic rocks
- Pangaea
- plate tectonics
- sedimentary rocks
- weathering

Rocks

There are three main types of rocks: sedimentary, metamorphic, and igneous. The differences among them are due to how they were formed. **Sedimentary rocks** are made of fragments of material such as sand, shells, and pebbles. These particles are often called sediments. They form as they are hardened under pressure over time. Sedimentary rocks are usually the only type that might contain fossils. Limestone is a sedimentary rock.

Igneous rocks form when magma (molten rock from within Earth) cools and hardens. Sometimes magma cools inside Earth, and sometimes it erupts as lava from a volcano and then cools. If the lava cools very quickly, it creates shiny, glass-like rocks. Obsidian is an example of an igneous rock.

Metamorphic rocks form under the surface of Earth, where other types of rock are hardened under heat and pressure. The composition of metamorphic rocks changes under pressure; the pressure causes them to develop ribbon-like layers or crystals. Marble is an example of metamorphic rock.

Fossils

Fossils are naturally preserved remains or traces of animal or plant life from the past. The two types of fossils are body fossils and trace fossils. A body fossil is the actual body of an animal or plant that has been preserved in nature. Some minerals, such as salt, enter the body of a living organism after death and prevent it from decomposing. Hard parts of a living organism, such as the bone structure or shell, are more resistant to decomposition. A trace fossil is only the evidence of the presence of a living organism. An example of a trace fossil is a rock with the shape of a fish imprinted naturally on it. The imprint is then preserved through natural processes similar to how a body fossil is preserved.

Body fossil of a nautilus

Changes in Earth

Continental Drift

Continental drift is a theory that describes how the continents moved on the surface of Earth to their current positions. The theory explains how similar fossils or similar animal and plant species are found on different continents. According to this theory, all the continents were joined together in the past but then split and moved to their current positions. Later, parts of the continental drift theory were rejected by geologists. The idea of a supercontinent, named Pangaea (which means "whole earth"), was confirmed. **Pangaea** (pan-JEE-uh), which existed about 200 million years ago, is believed to have been the last supercontinent on Earth.

The theory that replaced continental drift is called **plate tectonics**. Plate tectonics suggests that Earth's surface is made of rigid moving plates. These plates are larger than the continents and span different areas of Earth. This theory explains how various rocky plates glide over the inner layer of Earth. The cause of their movement is known to be due to the hot magma moving under the rocky plates. The movement of the plates is responsible for natural phenomena such as earthquakes and volcanic eruptions.

Tectonic Plates

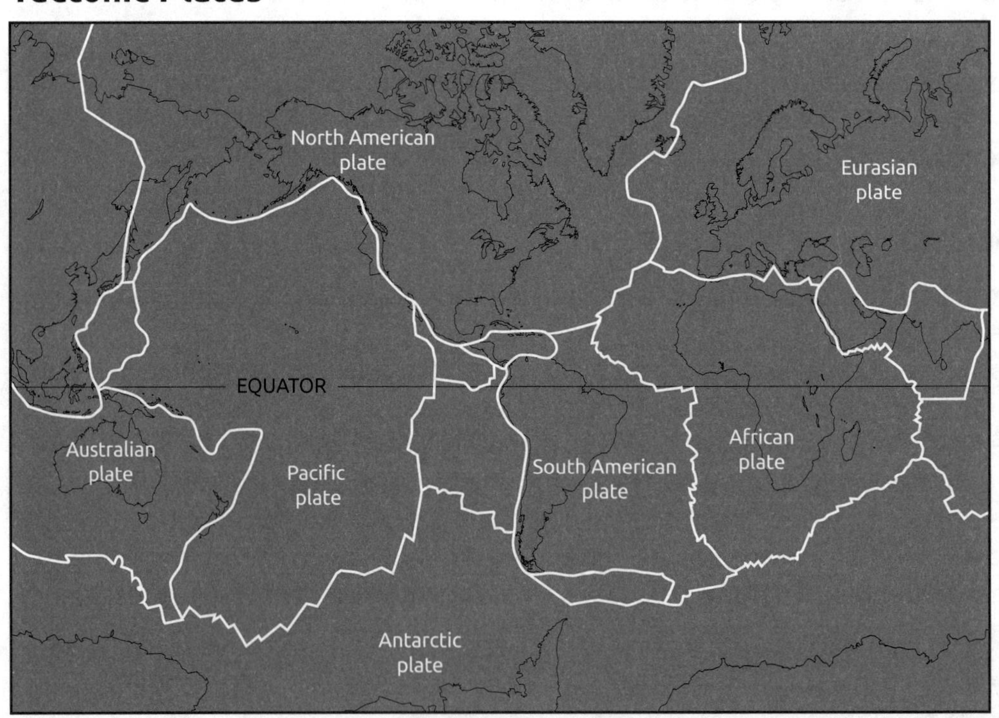

Weathering and Erosion

Weathering is the process of rock decomposition. Two processes are at work during weathering of rocks: chemical weathering and mechanical weathering. Chemical weathering involves a chemical reaction with some of the minerals of the rock. Chemical weathering impacts only the surface and is different from a chemical reaction taking place within the rock. Mechanical weathering is the physical process of rocks breaking into smaller pieces, leaving the chemical composition of the rock intact.

Changes in Earth

As a result of weathering, rocks break down into pieces and sometimes begin to move. As soon as rocks begin to move around, it is called **erosion**. Natural elements such as rain, ice, and wind cause the rocks to move around and flow. Falling rocks and debris moving down a slope due to gravity are examples of erosion. **Deposition** is the process of rocks and sediments moving to be deposited into a landform. The main difference between weathering and erosion is that weathering causes the rocks to loosen, while erosion is the process of movement of the rocks.

The processes of weathering, erosion, and deposition change the landscape of Earth. Wind and water, together with gravity, cause mountains to erode over time and become smaller. Riverbanks change shapes because of constant friction with water. While old rocks wash away, their sediments create a new landscape elsewhere. Weathering, erosion, and deposition are constantly destroying and creating landforms.

Lesson Practice

UNIT 2 / LESSON 3

Complete the activities below to check your understanding of the lesson content.

Skills Practice

Answer the questions based on the content covered in the lesson.

1. All of the following are examples of questions geologists seek to answer except which one?
   A. how tectonic plates move around
   B. why rocks break down into smaller pieces
   C. why the population of whales is decreasing
   D. what fossils can tell us about the history of Earth

2. Which of the following indicates a type of body fossil?
   A. the shape of a scallop shell on a rock
   B. the footprint of a mammoth in ice
   C. the trace of a fern on a crystal
   D. the tooth of a possum in saltwater

3. Explain the difference between the theories of plate tectonics and continental drift.

   ____

   ____

   ____

TEST STRATEGY

Underline the key words and phrases in the question and passage. For items with a passage, underline the key words and phrases in the question and then look for them in the passage.

KEY POINT!

The two types of fossils are body fossils and trace fossils.

Lesson Practice

KEY POINT!

The main difference between weathering and erosion is that weathering causes the rocks to loosen, while erosion is the process of movement of the rocks.

4. Explain the difference between chemical weathering and mechanical weathering.

5. All of the following elements cause erosion directly except which one?
   A rain
   B sunlight
   C snow
   D wind

6. Which type of rock is created when lava erupts from a volcano, then cools and hardens?
   A igneous
   B sedimentary
   C limestone
   D metamorphic

7. Which option is the correct description of Pangaea?
   A Pangaea is the theory that all continents were connected at once.
   B Pangaea is an alternative theory to plate tectonics.
   C Pangaea is the name of the last supercontinent on Earth.
   D Pangaea is the name of the largest tectonic plate on Earth.

8. Describe why a statue built of stone about 200 years ago is missing some details due to erosion.

See page 70 for answers and help.

Many elements work together to sustain life on Earth. The atmosphere, water, and soil are constantly exchanging matter. The movement of elements has evolved over millions of years to create balance. Recently, human activity has been changing some of these natural cycles. Since all elements of the environment are dependent on each other, imbalance in one part creates imbalance in all parts.

The Carbon Cycle

The **carbon cycle** is the movement of carbon through the environment. Carbon exists in the sea, in living matter, in the atmosphere, and even in rocks. Carbon is the building block of life and is an important part of many chemical processes. Carbon exists in the form of carbon dioxide in the atmosphere. Plants use carbon dioxide and sunlight to make food. Plants store carbon in their stems, leaves, and fruit. Animals eat the carbon in plants, and it becomes part of their bodies. When a plant or animal dies, carbon enters Earth and, under certain conditions, turns into fossil fuels. Fossil fuels are then burned by humans, and carbon is released in the form of carbon dioxide and returns to the atmosphere. This is one example of how carbon moves through its cycle on Earth. The following diagram shows some of the other ways carbon moves.

The Carbon Cycle

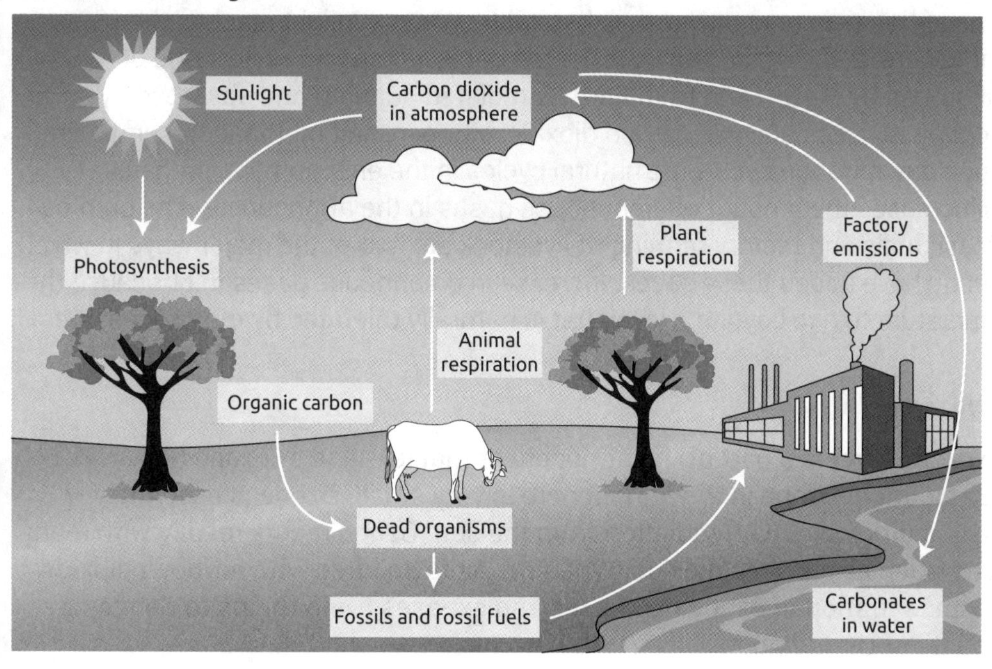

The Nitrogen Cycle

The **nitrogen cycle** is the movement of nitrogen through the environment. Nitrogen in the air and in the soil goes through many chemical processes and is constantly exchanged between the air and Earth. Plants require nitrogen to grow and produce seeds. The main source of nitrogen in soil is organic matter. Nitrogen in soil is not accessible by plants and has to be converted by a specific type of bacteria to be usable for plants. When the plant dies, the nitrogen stored in plants returns to the soil. The nitrogen in the air returns to the soil during rainstorms. The following diagram shows some of the other ways nitrogen moves.

KEY WORDS

- carbon cycle
- conservation
- greenhouse effect
- natural resources
- nitrogen cycle
- nonrenewable resources
- ozone layer
- pollution
- renewable resources

The Nitrogen Cycle

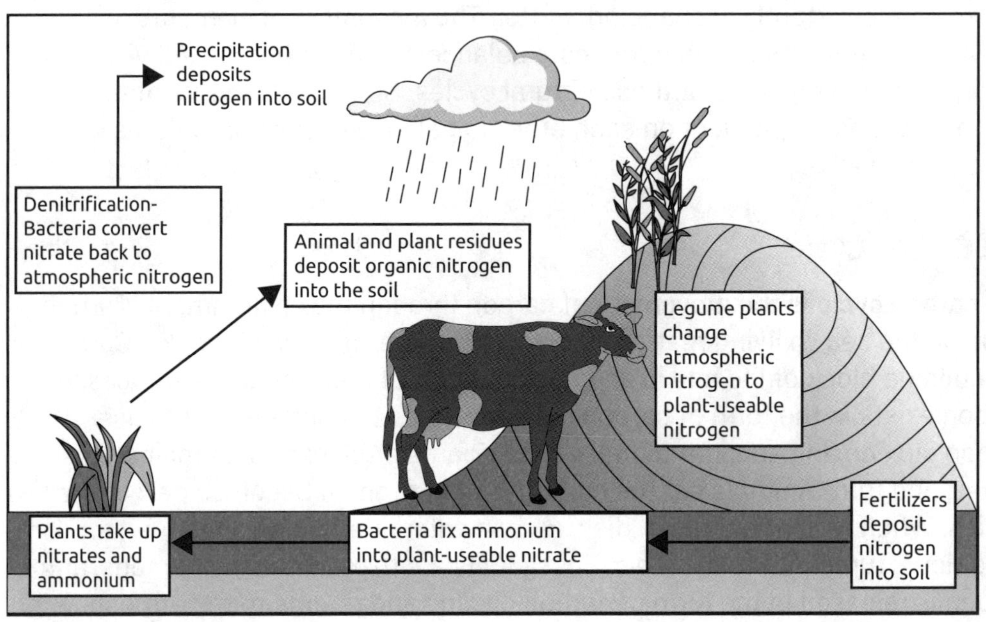

Greenhouse Effect

The atmosphere moderates the temperature on Earth. The **greenhouse effect** is one of the main reasons Earth is habitable. When sunlight reaches the atmosphere, some of the heat is reflected back into space, while some of it is absorbed by Earth. Heat from Earth is radiated outward and absorbed by greenhouse gases such as carbon dioxide, nitrogen, and methane. Whereas the carbon and nitrogen cycles are natural cycles in the environment, human activity has increased the amount of greenhouse gases in the atmosphere. The burning of fossil fuels and factory farming of livestock are two of the major ways in which humans have added these gases. Increase in greenhouse gases might cause the temperature to rise beyond a level that is naturally tolerated by many ecosystems.

Ozone Layer

The **ozone layer** is a part of the atmosphere that is made of ozone (O_3). The amount of ozone in the atmosphere is very small. Ozone is essential for blocking ultraviolet (UV) radiation from the sun. Ozone reacts readily with many gases, such as nitrogen dioxide. When the air is polluted with nitrogen dioxide, the concentration of ozone decreases and exposes living things to dangerous amounts of UV radiation.

Pollution

Pollution occurs when pollutants contaminate natural resources. **Natural resources** are the elements that sustain life on Earth. They include air, water, soil, and sources of energy. Pollution disturbs ecosystems and can lead to illness for humans and to extinction of some species. Air pollution is a dangerous and prominent form of pollution. Human activity releases various gases and particles into the atmosphere. These substances contribute to the greenhouse effect, the destruction of the ozone layer, and poor lung health in animals.

Water pollution is created when industrial waste enters our water sources. Water pollution can harm aquatic creatures and enter the food chain, thereby harming many other types of living beings. Soil pollution is the accumulation of toxic chemicals in soil due to human activity. Chemical pesticides, electronic waste, and deforestation are some of the factors that create pollution of soil.

Conservation

Conservation is the effort to protect and preserve natural resources and the lives sustained by them. Reducing pollution, creating biodiversity, and preserving nonrenewable resources are parts of that effort. **Nonrenewable resources** are resources that are being consumed faster than they are being produced. Fossil fuels such as coal, oil, and natural gas are examples of nonrenewable resources. Using **renewable resources** as a source of energy decreases pollution and creates a more sustainable way of life. Renewable resources are replaced quickly and can be used repeatedly. Sunlight, wind, and trees are examples of renewable resources.

KEY POINT!

Conservation is the effort made to protect and preserve natural resources and the lives sustained by them.

Lesson Practice

Complete the activities below to check your understanding of the lesson content.

Skills Practice

Answer the questions based on the content covered in the lesson.

1. Read the following passage.

   Chlorofluorocarbons, or CFCs, contain chlorine, fluorine, and carbon. CFCs are used in refrigerator insulation. CFCs are very stable until exposed to UV radiation. UV radiation breaks down CFCs and releases chlorine. Chlorine is extremely reactive with oxygen. One chlorine atom can remove thousands of ozone molecules from the atmosphere.

   Which statement is correct?

   A CFCs destroy thousands of ozone molecules because CFCs rise into the atmosphere.

   B Chlorine is released as a result of a chemical reaction between CFCs and ozone.

   C Oxygen gas is unstable and reacts to CFCs, which creates an imbalance that destroys ozone.

   D Sunlight breaks down CFCs and releases chlorine, which then destroys ozone molecules.

TEST STRATEGY

Look for any familiar terms and think about or jot down their meanings. For any unfamiliar terms, try to break down the words, using prefixes, suffixes, stems, and related forms of the words to determine meaning.

Lesson Practice

KEY POINT!

The ozone layer is essential for blocking the ultraviolet (UV) radiation from the sun.

2. Which natural process is described in the following statement?

   "Respiration releases carbon dioxide into the air. This gas is then used by plants during photosynthesis to make sugar and starch."

   A the carbon cycle
   B the nitrogen cycle
   C the greenhouse effect
   D conservation of biodiversity

3. Which of the following is an example of a renewable resource?

   A natural gas
   B paper
   C plastic
   D coal

4. Which of the following is NOT a conservation effort?

   A saving the lives of endangered species
   B switching to solar power for energy
   C using sunscreen to block UV radiation
   D reducing the release of greenhouse gases

5. Industrial waste runoff is a major factor in creating which kind of pollution?

   A water
   B soil
   C noise
   D air

6. Name and describe one way in which human activity creates air pollution.

   ____

   ____

7. Describe two ways that nitrogen enters the soil.

   ____

   ____

8. Briefly explain how the greenhouse effect helps keep Earth warm.

   ____

   ____

See page 70 for answers and help.

Theories about the Universe

The **big bang model** (often called the big bang theory) is a theory that explains the origin and evolution of our universe. It is widely accepted and is supported by decades of research and observation. According to the model, between 12 billion and 14 billion years ago, the observable universe was very small, about the size of a grain of sand, and very hot. It then expanded and cooled, but the hot dense matter left behind cosmic microwave background radiation. The detection of this radiation helped give scientists clues about the origin of the universe.

The matter in the universe is caught in a struggle between the momentum of expansion, caused by the push from the original hot, dense matter expanding, and the pull of **gravity**. Gravity is the force that causes all matter to be attracted to all other matter—the more matter, the stronger force. Current theories about gravity suggest that large amounts of matter distort nearby space. If the density of the universe is great enough, it would be positively curved, like a sphere, and therefore be finite, or closed, called the **closed universe theory**. If the density of the universe is small enough, then space would be negatively curved like the surface of a saddle and infinite, or referred to as the **open universe theory**. It turns out that the density of the universe is somewhere between too much and too little. The forces of gravity and momentum of expansion balance out, so the shape of the universe seems to be both flat and infinite. This is called the **flat universe theory**.

Lifecycle of Stars

Astronomy is the study of objects and formations in the universe, such as stars. Astronomers think that stars are formed in **nebulae**, dense clouds of gas in regions of relatively empty space. The matter in one part of a cloud will be brought together by gravity; when enough matter has collapsed together to form a central core that can fuse hydrogen to helium, it becomes a **main-sequence star**. Our sun is a main-sequence star. The more massive a main-sequence star, the brighter and bluer it is; less massive stars are dimmer and redder. A star's lifetime as a main-sequence star depends on the limited supply of hydrogen in its core and the rate at which it burns the hydrogen; larger stars burn faster than smaller stars.

Once a star about the size of the sun has used up its hydrogen, it starts to expand and becomes a **red giant**, which is redder and brighter. The outer part of the star continues to expand, but the core collapses due to gravity; when it is dense enough, the core will start converting helium to carbon. When all the helium is used up, the star will lose much of its mass as it becomes a hot core of carbon surrounded by ionized gases, called a **planetary nebula**. Eventually, the core will cool and become first a **white dwarf** and then a **black dwarf**.

KEY WORDS

- astronomy
- big bang model
- black dwarf
- black hole
- closed universe theory
- elliptical galaxy
- flat universe theory
- globular clusters
- gravity
- irregular galaxy
- main-sequence star
- Milky Way
- nebula(e)
- neutron star
- open universe theory
- planetary nebula
- red giant
- red supergiant
- solar system
- spiral galaxy
- supernova
- white dwarf

Stars that are much more massive than the sun also expand and become red **supergiants**; as they collapse, their carbon core remains massive enough to continue fusion, forming heavier elements from the carbon: oxygen, neon, silicon, sulfur, ending with iron. The iron core collapses to become a hot neutron core and then explodes, forming a **supernova**. The hot neutron core can then cool to form a **neutron star** or, if the original star was very massive, collapse further to form a **black hole**.

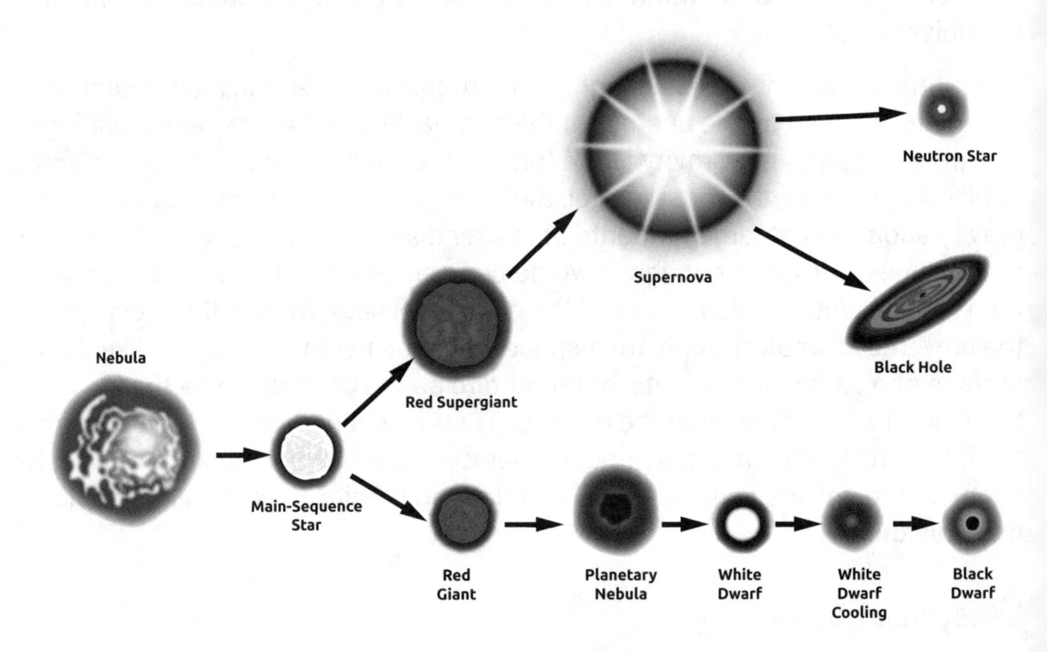

Galaxies and Solar Systems

Galaxies are groups of stars held together by gravity. Scientists have identified three types of galaxies, classifying them by their shape and appearance. **Elliptical galaxies** look circular or oval. **Spiral galaxies** have a center of older stars with "arms" of dust, gas, and younger stars spiraling outward from the center. They also have old clusters of stars, known as **globular clusters**, arranged in a loosely spherical shape around the center and arms. **Irregular galaxies** have no regular shape and can be made up of young stars, gas, and dust. Our galaxy, the **Milky Way**, is a spiral galaxy, and our solar system is located in one of its arms. **Solar systems** contain a star, planets, and other formations. The material that was around the star in the nebula as it formed is caught in the star's gravity; the material then clumps together as it moves around the star. Some of the material might have come from exploding stars nearby, which could also have started the star forming and provided some of the heavier elements for the planets and other formations, such as the iron found in Earth's crust.

Complete the activities below to check your understanding of the lesson content.

Skills Practice

Answer the questions based on the content covered in the lesson.

1. A scientist at NASA announced that one of its telescopes in space had discovered a new black hole. Which of the following could the new black hole have been during its lifetime?
   A a neutron star
   B a red supergiant
   C a black dwarf
   D a planetary nebula

2. One theory about how the universe might end is called the big crunch. It suggests that the expansion of the universe will slow down because the density is high enough for gravity to overcome the expansion. Once the expansion stops, the universe will start to collapse again, until all of the mass has come together. Which theory would support the idea of the big crunch?
   A big bang theory
   B open universe theory
   C closed universe theory
   D flat universe theory

3. Which option correctly lists the terms from smallest to largest in size?
   A solar system, star, universe, galaxy
   B universe, galaxy, star, solar system
   C galaxy, universe, solar system, star
   D star, solar system, galaxy, universe

4. An astronomer is looking at a galaxy that is similar to ours. Which type is that galaxy likely to be?
   A spiral
   B elliptical
   C globular
   D irregular

TEST STRATEGY

Rewrite the question, putting it in your own words to make sure you understand it. In the case of Question 1: "Which can produce a black hole: a neutron star, a red supergiant, a black dwarf, or a planetary nebula?"

KEY POINT!

Theories about the shape of the universe involve density being high, low, or "just right."

Lesson Practice

KEY POINT!

The big bang model states that the universe began between 12 billion and 14 billion years ago.

5. NASA has telescopes searching for background radiation throughout the observable universe. Are they likely to find radiation that is 15 billion years old or older? Why or why not?

____

____

____

____

____

____

6. Nebulae are also called "the birthplace of stars." Explain why they have that name.

____

____

____

____

____

____

See page 70 for answers and help.

During a war in the area now known as Turkey, in 585 BCE, two ancient kingdoms halted a battle when the sky darkened, the sun disappeared behind the moon, and nighttime seemed to come in the middle of the day. Unable to explain the phenomenon, the soldiers took it as a sign to end the war. Since astronomers can calculate exactly when the eclipse they saw occurred, the Battle of the Eclipse is perhaps the earliest historical event whose date is known with precision.

KEY WORDS

- heliocentric
- lunar eclipse
- penumbra
- phase
- solar eclipse
- sunspots
- tide
- umbra

The Moon and Its Phases

Over the course of a 29.5-day period, the moon goes through a cycle of **phases**, or the way the moon appears to an observer on Earth. The phases are caused by the revolution of the moon around Earth; as it moves around Earth, we see different portions of the moon that are illuminated by the sun.

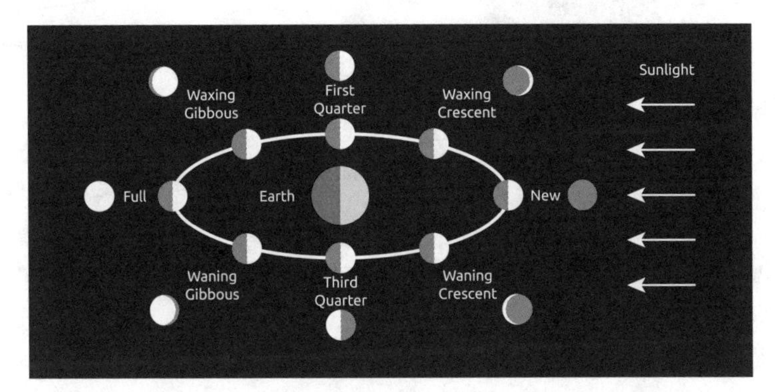

Phases of the moon as seen from above Earth (on the circle) and from the surface of Earth (outside the circle)

Eclipses

During the Battle of the Eclipse in 585 BCE, the soldiers witnessed a solar eclipse. In a **solar eclipse**, the moon passes directly between the sun and Earth. The moon obstructs the light from the sun and casts two circular shadows onto Earth, one inside the other. These two shadows are called the **penumbra**, within which a partial solar eclipse can be seen, and the smaller **umbra**, within which a total solar eclipse can be seen. A total eclipse (when the sun is completely blocked out) is visible only from the very small area of Earth's surface that falls within the umbra. A partial solar eclipse is visible to an observer who is within the penumbral shadow; this happens either if the observer is outside of the umbra during a total solar eclipse or if *only* the penumbra is cast on Earth because the moon, Earth, and sun are not perfectly lined up.

Solar Eclipse

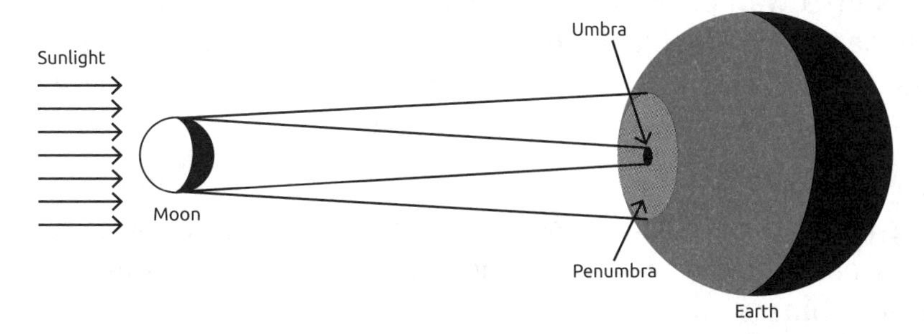

A solar eclipse occurs when the moon casts a shadow on Earth.

When the moon passes through Earth's shadow, it is no longer illuminated by the sun, and a **lunar eclipse** is observed; in this case, however, both the penumbra and umbra are large enough to cover the surface of the moon. When the moon passes through the center of Earth's umbra, a total lunar eclipse is seen from Earth. If only part of the moon travels through the umbra, or if the moon passes through just the penumbra, then a partial lunar eclipse is seen.

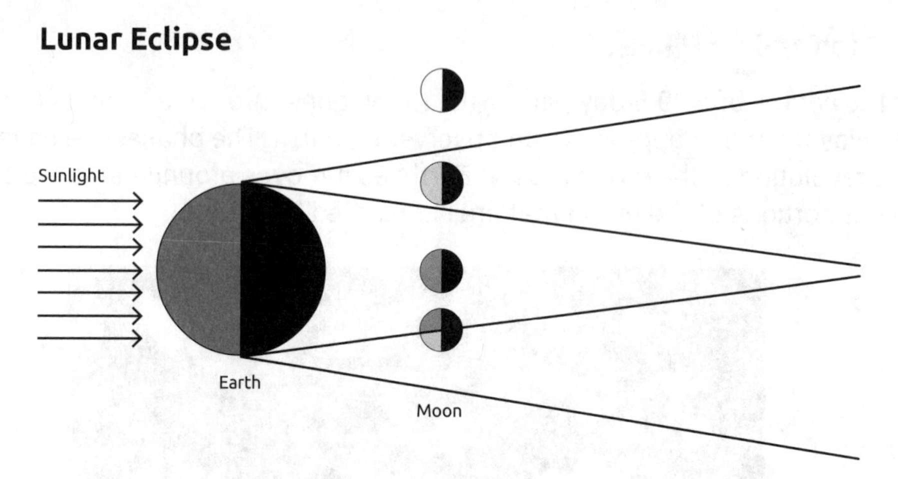

A lunar eclipse occurs when the moon passes through Earth's shadow.

Tides

As Earth rotates, the moon's gravity and the motion of Earth cause the oceans to bulge out in places; this phenomenon is known as **tides**. High tide occurs on the side of Earth facing the moon as well as on the opposite side, while the sides of Earth perpendicular to the moon experience low tide. Neap tides, the lowest high tides and highest low tides, occur during the first- and third-quarter phases of the moon, while spring tides, the highest high tides and lowest low tides, happen at full or new moon.

The Sun, Our Star

We did not always know, as we do now, that all the objects in our solar system revolve around the sun. Known as a **heliocentric** system, this theory was first introduced by Nicolaus Copernicus in the early sixteenth century. Before Copernicus, most people believed that the solar system was geocentric—that Earth was the center of the solar system, with all objects orbiting it.

The bodies revolving around the sun include the rocky inner planets, the gaseous outer planets, dwarf planets, satellites (moons), comets, asteroids, and meteors. Despite the number of objects that orbit it, the sun contains 99.8% of the mass in the solar system, and therefore has an immense gravitational pull. This keeps all the objects around it from leaving their orbits.

The sun does not have a solid surface, but it does have an atmosphere, which contains features including **sunspots**, or concentrations of the sun's magnetic field. Sunspots and their related phenomena can cause disruptions to satellites and power plants.

Complete the activities below to check your understanding of the lesson content.

Skills Practice

Answer the questions based on the content covered in the lesson.

1. During which of these moon phases does a neap tide occur?
   A full moon
   B third quarter
   C waxing crescent
   D waning gibbous

2. Which is the most massive object in our solar system?
   A Earth
   B comets
   C the sun
   D the moon

3. Which location in the diagram will experience a total lunar eclipse?

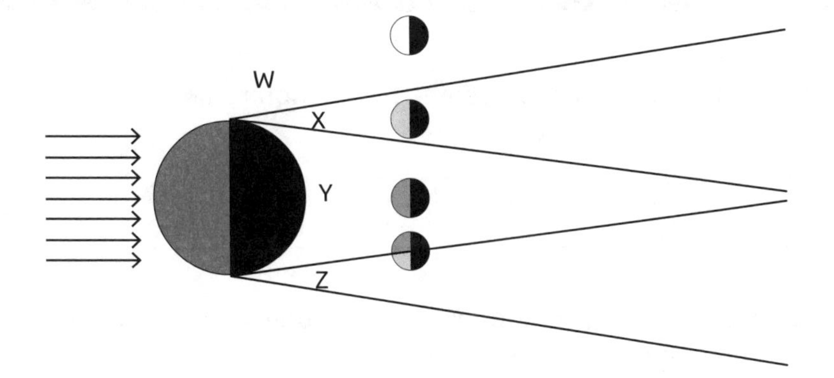

   A W
   B X
   C Y
   D Z

4. In which part of the moon's shadow must an observer be to experience a total solar eclipse?

   ____

5. Sunspots are located in the sun's ____________________.

6. During the ____________________ phase, no parts of the moon are visible to an observer on Earth.

TEST STRATEGY

Cover the answers to the question. Think of what you would write if the answers weren't there, and then uncover the answers. Choose the one that best matches your own answer.

Lesson Practice

KEY POINT!

Earth's shadow is made up of two parts: the umbra and the penumbra. During a lunar eclipse, observers in the umbra will see a total eclipse.

Use the following diagram to answer questions 7 and 8.

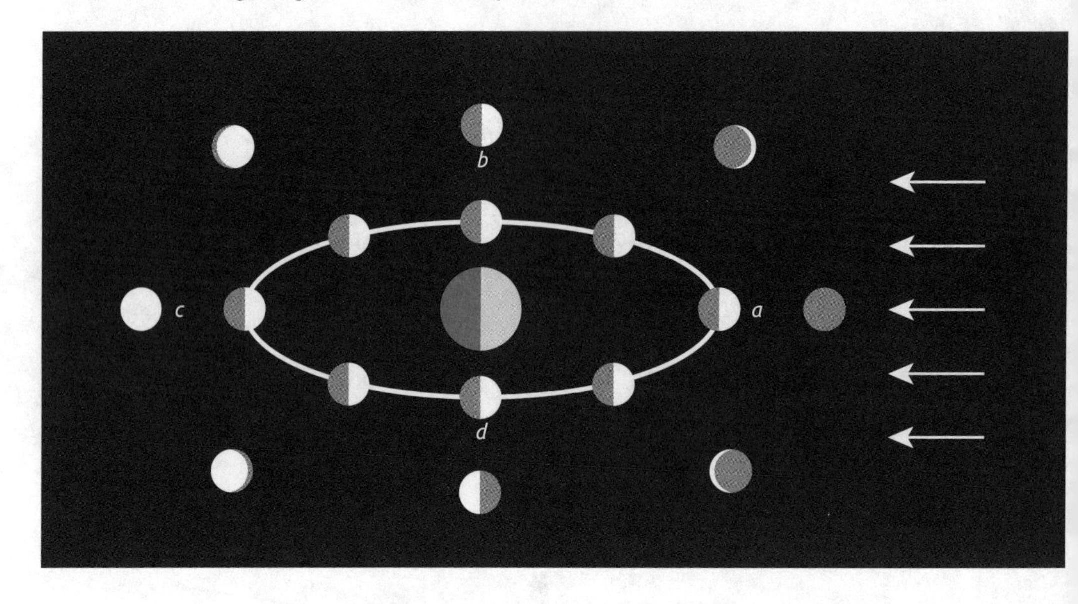

7. Which moon phases are represented by locations *a* and *b*?

   a: ______________________________

   b: ______________________________

8. The newspaper reports that today is a neap tide. At which location(s) on the diagram could the moon be in today?

   ____

See page 71 for answers and help.

Answer the questions based on the content covered in this unit.

Base your answers to questions 1–3 on the following passage and the unit content.

Hydraulic fracturing (fracking) is a method used to obtain fossil fuels, such as oil or natural gas, from underground reservoirs. The fossil fuels are usually trapped in porous rock such as sandstone or limestone, kept in the rock by the high pressures around them. To release the fossil fuels, workers drill between one and four miles down from the surface into the reservoir. They then pump large quantities of a water solution or fracking fluid to fracture the surrounding rock to make pathways for the fossil fuels to escape. The freed fossil fuels are then pumped to the surface. Some of the fracking fluid is also pumped to the surface and kept in storage ponds, but less than half of it is recovered. A typical well will use between three and eight million gallons of water in its lifetime. Environmentalists are concerned that the fracking fluid, which contains a variety of chemicals, will leak from the ponds and the underground reservoirs into surrounding areas. They are also concerned that methane, a greenhouse gas, can leak from the wells during drilling.

Another concern about fracking is that it can cause changes in stable faults, leading to earthquakes. The number of fracking wells has increased dramatically since 2000, after improvements in the technique made it profitable. Earthquakes in parts of the central and eastern United States have been blamed on fracking, as these areas are not near plate boundaries. Therefore, these areas expect to have few earthquakes above 3.0 on the Richter scale. Under normal circumstances, the stress on the stable faults in these areas is not enough to make them move. However, the additional weight of the fracking fluid above a fault and the change in friction caused by the fracking fluid wetting the rock can cause the faces of the fault to move past each other, triggering an earthquake.

1. In which layer of the Earth does fracking take place?
   - **A** crust
   - **B** inner core
   - **C** mantle
   - **D** outer core

2. Why is the leakage of methane from fracking a concern for environmentalists?
   - **A** Methane traps radiation from the sun, contributing to global warming.
   - **B** Methane leaches into groundwater, contaminating drinking water for humans.
   - **C** Methane increases the density of ocean water, disrupting global ocean currents.
   - **D** Methane replaces oxygen in the atmosphere, causing health problems for animals.

3. Fracking occurs in many areas in which knowledge of fault location is incomplete. How can this pose a problem to scientists?
   - **A** It can cause earthquakes to occur at a deeper focus than originally believed.
   - **B** It can make it difficult to predict areas in which injection of fluids could increase the risk of earthquakes.
   - **C** It can make it difficult to determine the most productive places to extract oil from surrounding rock.
   - **D** It can cause sedimentary rocks to change to metamorphic rocks due to intense heat and pressure.

Base your answers to questions 4 and 5 on the following diagram and the unit content.

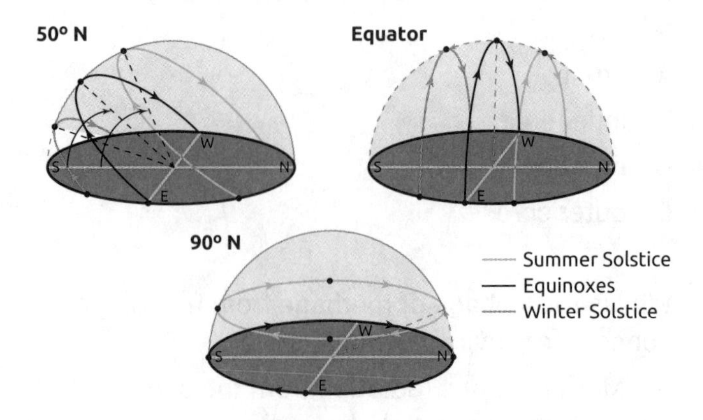

The diagrams show the sun's path through the sky at different locations on Earth.

4. Why is the sun's path through the sky highest during the summer solstice at 50°N?
   - **A** The earth is closest to the sun.
   - **B** The sun is stronger during the summer.
   - **C** The earth's revolution is faster during the summer.
   - **D** The Northern Hemisphere is tilted toward the sun.

5. Which statement most accurately describes the sun's path through the sky at 90°N?
   - **A** During the summer solstice, the sun provides 24 hours of daylight.
   - **B** During the summer solstice, the sun is not visible, causing 24 hours of darkness.
   - **C** During the summer solstice, the sun remains at the horizon.
   - **D** During the summer solstice, the sun rises in the east and sets in the west.

6. Air at the top of a mountain is often referred to as "thin" when compared to the air at sea level. Which of the following is directly related to this difference in density?
   - **A** The pressure of gases decreases with elevation.
   - **B** The composition of gases changes with elevation.
   - **C** The temperature of gases decreases with elevation.
   - **D** The condensation of gases occurs more rapidly with elevation.

Base your answer to question 7 on the following diagram and the unit content.

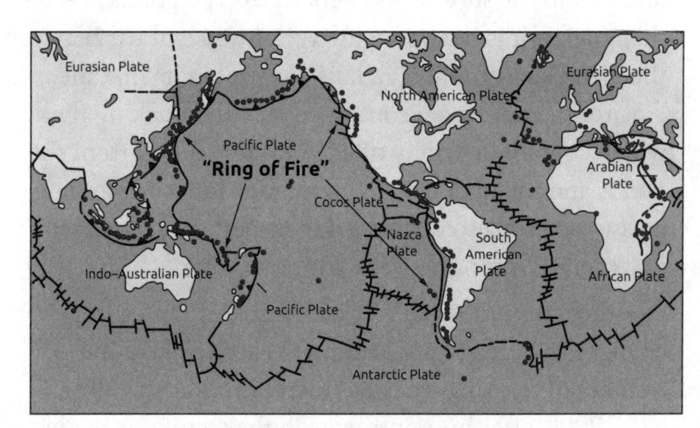

The diagram shows Earth's tectonic plates.

7. Why are most of the earth's active volcanoes along "the Ring of Fire?"
   - **A** "The Ring of Fire" is on plate boundaries.
   - **B** "The Ring of Fire" is hotter than other locations.
   - **C** "The Ring of Fire" is an area of strong air currents.
   - **D** "The Ring of Fire" is an area of strong ocean currents.

8. About an hour before Hurricane Sandy made landfall in New Jersey in 2012, it was reclassified as a "post-tropical cyclone." Which statement explains the reclassification of the storm?
   A It lost intensity as it moved over land.
   B It moved too far north to be called a hurricane.
   C It became a low-pressure system with high winds.
   D It received its energy from temperature differences in the atmosphere.

9. In a subduction zone, oceanic crust is pushed under continental crust, resulting in immense pressure and heat. Which type of rock is likely formed at a subduction zone?
   A igneous
   B metamorphic
   C sedimentary
   D volcanic

10. Which landform is most likely formed by mechanical and chemical weathering?
   A mountains
   B limestone caves
   C sandbars
   D soil

11. A bend in a river is called a *meander*. The water on the outside of a meander moves faster than on the inside of a meander. Over time, a meander can become separated from the rest of the river in what is known as an oxbow lake. Which statement best explains how an oxbow lake is formed?

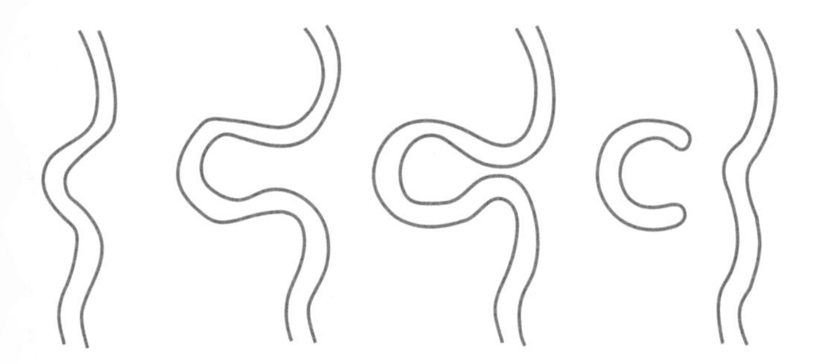

   A Erosion happens on the outside of a meander and deposition occurs on the inside of a meander until the neck of the meander is narrowed.
   B Deposition occurs on the outside of a meander and erosion occurs on the inside of a meander until the bank of the river is changed.
   C Erosion occurs on both sides of the meander until the rate of movement of the water in the river is changed so that it becomes faster on the inside of the meander.
   D Deposition occurs on both sides of the meander until the river becomes completely filled in and water can no longer move through its banks.

12. Deforestation is a process in which forests, especially tropical rain forests, are destroyed for human purposes. Two methods of deforestation are clear cutting and burning.
   All of the following are negative effects of deforestation EXCEPT:
   A increase of carbon dioxide in the atmosphere
   B destruction of habitat
   C halting of the nitrogen cycle in that location
   D loss of a nonrenewable resource

13. A star has used up all of its hydrogen. What is likely to happen next?
   A It will begin using helium to produce more hydrogen.
   B It will begin using carbon to produce more hydrogen.
   C It will begin using helium to produce carbon.
   D It will begin using hydrogen from nearby nebulae.

14. Which is the best description of a galaxy?
   A a collection of planets
   B a collection of nebulae
   C a collection of stars
   D a collection of dense gases

15. Which likely had the most effect on the shape of the universe?
    - **A** the density and the rate of expansion
    - **B** the amount of cosmic background radiation
    - **C** the age of the stars near the center of the universe
    - **D** the amount of matter that existed before the big bang occurred

16. A very massive supergiant explodes. What is it likely to become?
    - **A** a neutron star
    - **B** a red supergiant
    - **C** a nebula
    - **D** a black hole

17. Why are high tides higher during spring?
    - **A** The gravitational pull of the sun is added to the gravitational pull of the moon.
    - **B** The gravitational pull of the sun partially cancels out the gravitational pull of the moon.
    - **C** The earth's center of gravity pushes the water out farther, towards the gravitational pull of the moon.
    - **D** The moon is closer to the earth in spring, thus increasing the moon's gravitational pull on the earth.

Base your answer to question 18 on the following diagram and the unit content.

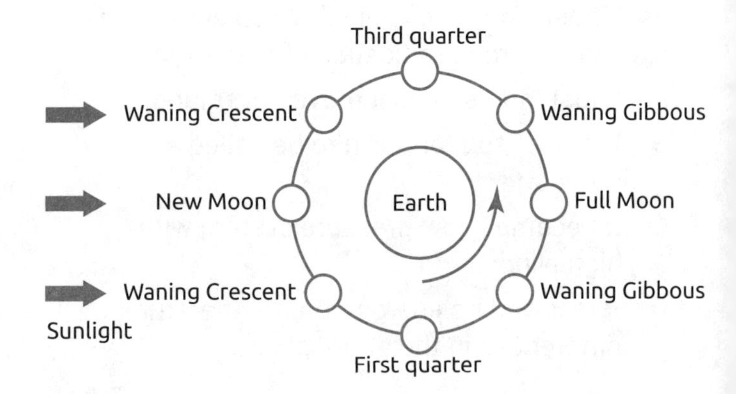

18. Which two events could occur in the same 24-hour period?
    - **A** waning crescent and solar eclipse
    - **B** new moon and solar eclipse
    - **C** first quarter and lunar eclipse
    - **D** waxing gibbous and lunar eclipse

Base your answers to questions 19–21 on the following passage and the unit content.

On March 6, 2009, a rocket carrying NASA's Kepler spacecraft lifted from Cape Canaveral Air Force Station in Florida. Kepler was heading on a three-and-a-half-year mission to search for signs of other Earth-like planets. The Kepler spacecraft watched an area in space that contains about 150,000 stars that are similar to our sun. Its place in space allowed it to watch the same stars constantly throughout its mission, something ground-based telescopes cannot do. The spacecraft was in a heliocentric Earth-trailing orbit. This freed it from the gravitational and atmospheric disturbances it would experience by orbiting the Earth, but kept it close enough for communication. Kepler's telescope had special detectors, similar to those used in digital cameras, to detect very small changes in light. It used these to look for a slight dimming in the stars as planets pass between the stars and Kepler.

Kepler's mission ended in 2013, but scientists are still examining the data collected. On July 23, 2015, NASA announced that Kepler had found a planet, about the same size as Earth, in the "habitable zone" around a G2-type star. The habitable zone is the area around a star where liquid water can condense on the surface of an orbiting planet. Named Kepler-452b, the planet is 60 percent larger in diameter than Earth and has an orbit of 385 days. Its star, which is the same type of star as our sun, is larger and brighter than our sun but has the same temperature. Kepler-452b is about 5 percent farther away from its star than Earth is from our sun and is about 1.5 billion years older than Earth. It is quite possible that Kepler-452b will be one of many Earth-like planets observed during the mission.

See page 71 for answers.

19. Which of these best describes Kepler-542b's star?
    - **A** neutron star
    - **B** main-sequence star
    - **C** red giant
    - **D** white dwarf

20. Kepler-542b's star is likely to have a shorter lifetime than Earth's sun. Which of these is the best explanation for the shorter lifespan?
    - **A** It was formed about 1.5 billion years before Earth's sun.
    - **B** Its planets have larger orbits than those around Earth's sun.
    - **C** It is larger and brighter than Earth's sun.
    - **D** Its planets are larger than those around Earth's sun.

21. How did the Kepler spacecraft travel in space?
    - **A** It revolved around the sun.
    - **B** It revolved around Earth.
    - **C** It traveled toward the stars it was investigating.
    - **D** It traveled on the same orbit as Earth but in the opposite direction.

Lesson 1

1. **summer.** The Northern Hemisphere receives more direct sunlight when it is tilted toward the sun.
2. **Possible answer: The metallic, liquid outer core spins as Earth rotates.**
3. **exosphere.** The exosphere is the outermost layer of Earth's atmosphere, where the gases have low enough density to not harm or slow satellites too much.
4. **D.** Weather occurs in the atmosphere's lowest layer.
5. **B.** The lithosphere is mostly solid rock made up of the crust and upper mantle.
6. **A.** The continents are part of Earth's outer rocky layer—the crust—and the oceans sit atop that layer.
7. **C.** Rotation causes day and night, so if rotation were to cease, then one side of Earth would be in constant daylight while the opposite side would be in constant darkness.
8. **D.** Because of Earth's tilt, the sun's light falls at different angles on Earth depending on where Earth is in its yearly orbit. This causes the changing seasons.

Lesson 2

1. **cold**
2. **Gulf Stream**
3. **Possible answer: An earthquake's energy travels through Earth's interior by seismic waves, and the waves can reach places far away from the earthquake's focus.**
4. **earthquake, underwater landslide**
5. **A.** Several hours of light to moderate rainfall is typical weather before a warm front.
6. **C.** Tsunamis are caused by earthquakes and underwater landslides triggered by earthquakes. Earthquakes usually occur along plate boundaries, and the East Coast is not near a major plate boundary.
7. **B.** Large earthquakes occur in areas of infrequent seismic activity. They build up energy for many years and then suddenly release it.
8. **A.** As warm, moist air rises into the atmosphere, it condenses and forms cumulonimbus clouds heavy with moisture.
9. **A.** The southeastern United States has a lot of warm, humid weather, perfect for the generation of thunderstorms.
10. **A.** Hurricanes need warm water to form and to be sustained.

Lesson 3

1. **C.** Finding answers about the population of whales is a question for marine biologists. Geologists engage in questions about rocks, fossils, and the movement of tectonic plates.
2. **D.**
3. **Possible answer: Continental drift was a theory that described how the continents moved on the surface of Earth to their current position. Plate tectonics suggests that Earth's surface is made of rigid, moving plates. These plates are different from the continents and span different areas of Earth.**
4. **Possible answer: Chemical weathering involves a chemical reaction with some of the minerals of the rock. Mechanical weathering is the physical process of rocks breaking into smaller pieces.**
5. **B.**
6. **A.**
7. **C.**
8. **Possible answer: A statue built of stone starts to be impacted by weathering. Wind, rain, and snow are some of the natural elements that loosen the structure of the stone. Then, gravity causes the loosened pieces to fall.**

Lesson 4

1. **D.**
2. **A.**
3. **B.**
4. **C.**
5. **A.**
6. **Possible answer: releasing carbon dioxide into the air by burning fossil fuels**
7. **Possible answer: Rainstorms deliver atmospheric nitrogen to the soil. The nitrogen in plants and animals returns to the soil when those organisms die.**
8. **Possible answer: When sunlight reaches the atmosphere, some of it is reflected back into space, while some of it is absorbed by Earth. Heat from Earth is radiated outward and absorbed by greenhouse gases, such as carbon dioxide, nitrogen, and methane. The greenhouse gases prevent heat from escaping into space.**

Lesson 5

1. **B.**
2. **C.**
3. **D.**
4. **A.**

5. Possible answer: They are unlikely to find radiation that old, as the universe is between 12 and 14 billion years old; as far as we know, nothing existed to give off radiation before that.

6. Possible answer: Stars are formed when the dense gases in a nebula begin to clump together. Then, the increased gravity attracts still more material until the core is dense enough to start fusing hydrogen into helium, forming a star.

Lesson 6

1. B. A neap tide occurs when the sun, moon, and Earth are at right angles to each other, which is during the first and third quarters.

2. C. The sun contains 98.8% of all the mass in our solar system.

3. C. A total lunar eclipse occurs in areas inside the umbral shadow.

4. umbra A total solar eclipse is observed in the inner part of the shadow, or the umbra.

5. atmosphere. Observable features of the sun, such as sunspots, are located in the sun's atmosphere.

6. new moon During a new moon, the illuminated portion of the moon is facing away from Earth.

7 a: new, b: first quarter Location *a* is the new moon, and location *b* is the first-quarter phase.

8. a or c A neap tide occurs during new moon and full moon phases.

Unit Test

1. A.
2. A.
3. B.
4. D.
5. A.
6. A.
7. A.
8. D.
9. B.
10. B.
11. A.
12. C.
13. C.
14. C.
15. A.
16. D.
17. A.
18. B.
19. B.
20. C.
21. A.

Unit Glossary

- **air mass** – a large body of air with uniform characteristics
- **asthenosphere** – a layer of the Earth that lies about 50 miles below the surface
- **astronomy** – the study of objects and formations in the universe
- **atmosphere** – an envelope of gases surrounding Earth
- **axis** – an imaginary line, from north to south, through the center of a planet
- **big bang model** – theory that explains the origin of our universe
- **black dwarf** – the stage of a star's life following white dwarf
- **black hole** – the end stage of a very massive star's life; follows supernova
- **carbon cycle** – the movement of carbon through the environment
- **closed universe theory** – theory suggesting that if the universe has a great density, it is closed like a sphere
- **conservation** – the effort to protect and preserve natural resources
- **continental drift** – a theory that describes how the continents moved on the surface of Earth to their current position
- **core** – the innermost layer of Earth; consists of an outer core and an inner core

Unit Glossary

- **crust** – the outermost layer of Earth
- **current** – a body of water moving in a direction
- **deposition** – the process of rocks and sediments being deposited
- **earthquake** – an event resulting from the shifting of tectonic plates
- **elliptical galaxy** – a galaxy that looks circular or oval
- **erosion** – movement of weathered rock from one place to another
- **exosphere** – the highest part of Earth's atmosphere
- **flat universe theory** – theory suggesting that the universe has a density that is neither great nor little and is flat and infinite
- **focus** – the area of energy release inside Earth causing an earthquake
- **fossils** – naturally preserved remains or traces of animal or plant life from the past
- **front** – the boundary where two air masses meet
- **globular clusters** – old clusters of stars within a spiral galaxy
- **gravity** – the force that causes all matter to be attracted to other matter
- **greenhouse effect** – the trapping of heat by gases in the atmosphere
- **heliocentric** – theory that all of the objects in our solar system revolve around the sun, introduced by Nicolaus Copernicus in the early 16th century
- **humidity** – the amount of moisture in the air
- **hurricane** – a large low-pressure system with organized circulation
- **hydrosphere** – the "sphere" that contains all of Earth's water
- **igneous rocks** – rocks formed when magma cools and hardens
- **irregular galaxy** – a galaxy with no regular shape
- **lithosphere** – the upper part of the mantle and the crust
- **lunar eclipse** – phenomenon that occurs when the moon passes through the Earth's shadow
- **magma** – rock that exists in a semi-solid state in the mantle
- **main-sequence star** – a star that has a central core that fuses hydrogen to form helium
- **mantle** – the semi-solid layer of Earth beneath the core
- **mesosphere** – the third highest layer of Earth's atmosphere
- **metamorphic rocks** – rocks formed when other rocks are subjected to heat and pressure
- **Milky Way** – our galaxy; a spiral galaxy
- **natural resources** – materials that occur in nature
- **nebula(e)** – dense clouds in regions of relatively empty space
- **neutron star** – the stage of a star's life following supernova
- **nitrogen cycle** – the movement of nitrogen through the environment
- **nonrenewable resources** – resources that are being consumed faster than they are produced
- **open universe theory** – theory suggesting that if the universe has a small density, it is curved negatively and is open and infinite
- **ozone layer** – part of the atmosphere made of ozone
- **Pangaea** – the last supercontinent on Earth
- **penumbra** – a shadow that is cast during a solar eclipse, a partial solar can be seen within this shadow
- **phase** – the way the moon appears to an observer on Earth; it occurs because we see different portions of the moon illuminated by the sun as the moon orbits around Earth
- **planetary nebula** – a red giant that has collapsed and becomes a hot core of carbon
- **plate tectonics** – the theory that describes Earth's surface being made of rigid, moving plates
- **pollution** – the contamination of natural resources
- **red giant** – a star that has used up its hydrogen and expands
- **red supergiant** – a star more massive than the sun that has used up its hydrogen
- **renewable resources** – resources that are replaced quickly and can be used repeatedly

Unit Glossary

- **revolution** – the path of an object around another object
- **Richter scale** – scale used to measure the magnitude of an earthquake
- **rotating** – the turning of an object on its axis
- **seasons** – spring, summer, autumn, winter
- **sedimentary rocks** – rocks made from fragments of material such as sand, shells, and pebbles
- **seismic waves** – waves that carry the energy released from an earthquake
- **solar eclipse** – the passing of the moon directly between the sun and Earth
- **solar system** – a system that contains a star, planets, and other objects
- **spiral galaxy** – a galaxy that has a center of old stars with arms of dust, gas, and younger stars
- **stratosphere** – the second highest layer of Earth's atmosphere
- **sunspots** – concentrations of the sun's magnetic field
- **supernova** – the explosion of a red supergiant
- **thermosphere** – the fourth highest layer of Earth's atmosphere
- **tide** – phenomenon that occurs when the moon's gravity and the Earth's rotation cause the oceans to bulge out in places
- **troposphere** – the lowest layer of Earth's atmosphere
- **tsunami** – a set of ocean waves triggered by an earthquake
- **umbra** – a smaller shadow that is cast during a solar eclipse, a total solar eclipse can be seen within this shadow
- **weathering** – the process of rock decomposition
- **white dwarf** – the stage of a star's life following planetary nebula

Study More!

Consider exploring these concepts, which were not introduced in the unit:

- glaciers
- subduction of tectonic plates
- fault lines
- evaporation and condensation in the water cycle
- humidity and relative humidity
- types of precipitation and how they form
- Coriolis effect
- wind abrasion and creep
- rock exfoliation
- soil leaching
- hazardous waste
- solid waste disposal
- recycling
- extinction prevention
- pulsars and quasars
- light-years
- composition of asteroids, comets, and meteors
- NASA and artificial satellites and orbiters

UNIT 3

Physical Science

A simple bike race can elicit many questions in the mind of a physical scientist: What chemical reactions are providing energy for the bikers? How does that energy get transferred into motion? Why is one metal better than others for making bike frames? How does the pressure in the tires change when they get warm? How much force do the bikers use to go up a hill?

KEY WORDS

- acceleration
- acid
- activation energy
- anion
- atom
- base
- batteries
- boiling point
- catalyst
- cation
- chemical bond
- chemical change
- chemical equation
- chemical formula
- chemical reaction
- circuit
- colloid
- combination reaction
- compound
- condense
- conduction
- convection
- covalent bond
- current
- decomposition reaction
- distance
- double bond

Physical Science

UNIT 3

KEY WORDS

- electricity
- electromagnetic spectrum
- electron
- electron cloud
- electrostatic attraction
- element
- endothermic reaction
- exothermic reaction
- fluid friction
- force
- freeze
- frequency
- friction
- fulcrum
- gas
- gravitational force
- heterogeneous
- homogeneous
- hydroelectric
- inclined plane
- ionic bond
- kinetic energy
- Law of Conservation of Energy
- Law of Conservation of Matter
- lever
- light energy
- liquid
- magnetic force
- mass
- melting point
- metalloids
- metals
- mixture
- molecule
- momentum
- net force
- neutron
- non-metals
- non-polar
- nuclear force
- nuclear power
- nucleus
- Periodic Law
- periodic table of elements
- pH scale
- physical change
- polar
- potential energy
- product
- proton
- pulley
- radiation
- reactant
- reaction rate
- rolling friction
- sliding friction
- solar
- solid
- solute
- solution
- solvent
- static friction
- suspension
- synthesis reaction
- unshared electrons
- velocity
- wavelength
- wind power

Structure of Atoms

KEY WORDS

- atom
- electron
- electron cloud
- element
- metalloids
- metals
- neutron
- non-metals
- nuclear force
- nucleus
- Periodic Law
- periodic table of elements
- proton

KEY POINT!

Atoms are the smallest unit of matter that still have the same properties as the element they came from. They can be further broken down into protons, electrons, and neutrons, but these will have different properties.

If you are making a cake and you don't have enough sugar, you might substitute honey or maple syrup. Your knowledge that all three of these things are sweet helps you make that decision. Likewise, a chemist's knowledge about the properties of elements and compounds helps him or her choose the right ones for a particular use. With over 100 elements and many thousands of compounds, knowing the properties of all of them would seem impossible. Fortunately, chemists have been able to organize much of this information in one easy-to-use tool: the periodic table.

Atomic Structure and the Periodic Table

A piece of aluminum consists entirely of aluminum **atoms**; an aluminum atom is the smallest unit in that piece that still has the properties of aluminum. Aluminum is an **element**; all elements contain atoms that have the properties of that element and are different from atoms of other elements. All atoms are made up of smaller particles, called **protons, electrons**, and **neutrons**. The protons and neutrons have about the same mass as each other and are found in the **nucleus**, or center, of the atom; they are held together by a very strong **nuclear force**. The electrons have almost no mass and are found outside the nucleus; they move very fast in areas called **electron clouds**. The protons carry a positive charge, the electrons a negative charge, and the neutrons carry no charge. Electrons stay in the electron cloud region because they are simultaneously attracted inward toward the positive nucleus and pushed outward by their kinetic energy.

All atoms of a particular element have the same number of protons in their nucleus. This number is also called the atomic number, and it is specific to each element. Atoms have no charge, which means that its number of electrons must equal its number of protons. A neutral atom of aluminum has 13 protons and 13 electrons. When the elements are arranged in a table, going from lowest atomic number to highest, elements in the same column have similar

Periodic Table of the Elements

Non-metal; Alkali metal; Alkaline earth metal; Transition metal; Metal; Metalloid; Halogen; Noble gas; Lanthanide; Actinide

1	2	3	4	5	6	7	8	9	10	11	12	13	14	15	16	17	18
1 H																	2 He
3 Li	4 Be											5 B	6 C	7 N	8 O	9 F	10 Ne
11 Na	12 Mg											13 Al	14 Si	15 P	16 S	17 Cl	18 Ar
19 K	20 Ca	21 Sc	22 Ti	23 V	24 Cr	25 Mn	26 Fe	27 Co	28 Ni	29 Cu	30 Zn	31 Ga	32 Ge	33 As	34 Se	35 Br	36 Kr
37 Rb	38 Sr	39 Y	40 Zr	41 Nb	42 Mo	43 Tc	44 Ru	45 Rh	46 Pd	47 Ag	48 Cd	49 In	50 Sn	51 Sb	52 Te	53 I	54 Xe
55 Cs	56 Ba	57-71*	72 Hf	73 Ta	74 W	75 Re	76 Os	77 Ir	78 Pt	79 Au	80 Hg	81 Tl	82 Pb	83 Bi	84 Po	85 At	86 Rn
87 Fr	88 Ra	89-103**	104 Rf	105 Db	106 Sg	107 Bh	108 Hs	109 Mt	110 Ds	111 Rg	112 Cn	113 Uut	114 Fl	115 Uup	116 Lv	117 Uus	118 Uuo

*	57 La	58 Ce	59 Pr	60 Nd	61 Pm	62 Sm	63 Eu	64 Gd	65 Tb	66 Dy	67 Ho	68 Er	69 Tm	70 Yb	71 Lu
**	89 Ac	90 Th	91 Pa	92 U	93 Np	94 Pu	95 Am	96 Cm	97 Bk	98 Cf	99 Es	100 Fm	101 Md	102 No	103 Lr

properties; this is due to a natural law called the **Periodic Law**. For the first 20 atomic numbers, only 8 columns are needed to accommodate the law. After that, more columns must be added to get the same effect. The result is the **periodic table of elements**.

The rows in the periodic table are called *periods*, and the columns are called *groups*; the elements in a group have similar properties. For example, elements in group 1 (Li, Na, etc.) react violently, producing heat and light, when put in water, while the elements in group 18 (He, Ne, Ar, etc.) tend to be inert and not react with anything. Some trends can be predicted in groups and periods; for example, atomic size decreases as you go left to right in a period and increases as you go down a group.

Several groups of elements also have some similar basic properties, based on their location in the periodic table. Divide the table into two parts along the line created by B, Si, Ge, As, Sb, and Te. All of the elements to the left of this line have the properties of **metals**: they conduct electricity and heat well, and they are malleable and shiny. All of the elements to the right of the line have opposite properties; they do not conduct electricity, and they are brittle and dull as solids (most are gases at room temperature). These elements are called **non-metals**. The elements that make up the dividing line have some properties of metals and some of non-metals and thus are called **metalloids**.

KEY POINT!

People are often surprised to learn that sodium is a metal because they are more familiar with it as a component of salt. Pure sodium, however, looks a lot like a piece of aluminum, but it is softer and much more reactive, so don't expect to see a ladder or a boat made from sodium!

Lesson Practice

TEST STRATEGY

Draw a diagram to represent the information. Draw five circles to represent the nuclei of the four atoms listed as choices and the one mentioned in the question. Write + signs inside each circle to represent the protons and – signs outside the circles to represent the electrons. Which element's sketch most closely resembles that of the one mentioned in the question?

KEY POINT!

There are many more metals on the periodic table than non-metals, and metals react differently when combined with other elements than non-metals do.

Complete the activities below to check your understanding of the lesson content.

Skills Practice

Answer the questions based on the content covered in the lesson.

1. Which element would be the best to use in making an electrical wire?
   A sulfur (S)
   B gold (Au)
   C iodine (I)
   D argon (Ar)

2. Which element would have properties most similar to those of fluorine (F)?
   A chlorine (Cl)
   B oxygen (O)
   C neon (Ne)
   D lithium (Li)

3. What properties would you expect sodium (Na) to have?
   A poor conductor, brittle, dull
   B poor conductor, malleable, shiny
   C good conductor, brittle, dull
   D good conductor, malleable, shiny

4. A mystery element conducts electricity and is shiny but brittle. Which of these elements could it be?
   A carbon (C)
   B copper (Cu)
   C silicon (Si)
   D potassium (K)

Base your answers to questions 5–7 on the following passage and on the content of this lesson.

A librarian consults a chemist to discuss the best way to protect an old, valuable book. They know they must keep oxygen and air pollutants away from the book, so they decide to enclose the book in an air-tight glass case and fill the case with an inert gas. They know that helium is inert, but the helium atoms are so small that they leak out through tiny cracks in the seals on the box. Rather than having to refill the box with helium so often, the chemist suggests using argon because it has larger atoms than helium.

5. Which statement best describes how the chemist knows argon is also inert?
   A It has the same number of electrons as helium.
   B It is in the same periodas helium.
   C It is in the same groupas helium.
   D It has the same number of neutrons as helium.

6. Which property does the chemist expect argon to have?
   A non-conductive
   B reactive
   C shiny
   D malleable

7. Which element might be even better to use than argon?
   A neon (Ne)
   B xenon (Xe)
   C chlorine (Cl)
   D hydrogen (H)

See page 114 for answers and help.

Molecules

KEY WORDS

- anion
- cation
- chemical bond
- chemical formula
- compound
- covalent bond
- double bond
- electrostatic attraction
- ionic bond
- molecule
- non-polar
- polar
- unshared electrons

KEY POINT!

Metals plus non-metals tend to form ionic compounds. Non-metals tend to form covalent bonds with each other.

Can you tell the difference between salt and sugar without tasting them? A chemist can. With a few simple tests, chemists can predict the properties of compounds based on their formulas and on seeing in what way the atoms have combined to form them.

Types of Chemical Bonds

Atoms can combine in a chemical reaction to form **compounds**, which are held together by **chemical bonds**. A chemical bond forms when electrons are either transferred between atoms or shared between atoms. When electrons are transferred, an ionic compound is formed. The atoms that lost electrons become positively charged **cations**, and the atoms that gained electrons become negatively charged **anions**. The strong **electrostatic attraction** between the positive and negative ions is called an **ionic bond**. For example, ionic bonds are formed when potassium (K) reacts with chlorine (Cl); potassium loses an electron, and chlorine gains an electron. The potassium cation has a +1 charge, and the chlorine anion has a –1 charge. Metals tend to lose electrons, while non-metals tend to gain them, so ionic compounds always consist of a metal and a non-metal.

We can predict how many electrons some elements will gain or lose based on their group. Group 1 metals always lose one electron, group 2 metals always lose two electrons, and group 17 non-metals always gain one electron. The charges of the ions must cancel out, so the ratio of the ions will depend on the charges. For example, calcium will lose two electrons, and fluoride will gain only one, so calcium fluoride needs two fluorine atoms for every calcium atom. We include this ratio in the **chemical formula**, which uses atom symbols and subscripts. Potassium chloride is KCl, because the ratio is 1:1. Calcium fluoride is CaF_2, because the ratio is 1:2.

Covalent bonds tend to occur between non-metals, where the atoms share electrons instead of transferring them. Hydrogen also tends to form covalent bonds with non-metals, even though it is on the left side of the periodic table. One bond is formed for every two electrons shared between atoms (represented by the dotted lines in the diagram). For example, if two hydrogen atoms react with one atom of oxygen, two covalent bonds are formed, one between each hydrogen and the oxygen atom. The formula for this compound (water) is H_2O.

More than one bond can form between two atoms. For example, in carbon dioxide (CO_2), carbon has a **double bond** with each oxygen molecule. A **molecule** is the basic unit for a compound, the way an atom is the basic unit for an element. Ionic compounds do not actually form molecules, as the electrostatic attraction is not just between individual ions.

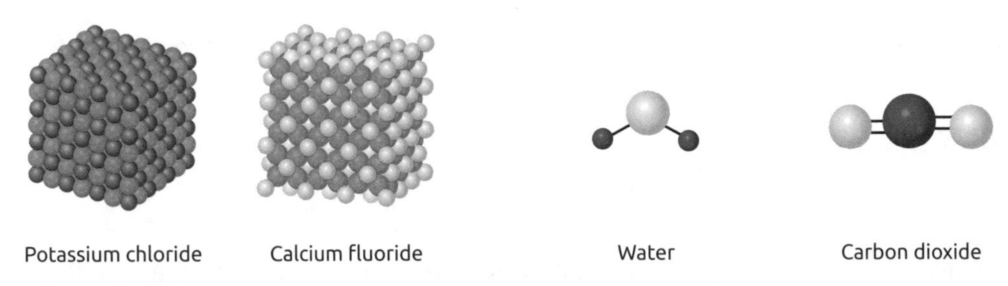

Representations of ionic and covalent compounds

Molecular Shapes

When elements make covalent bonds, they share some or all of their outermost electrons. Any **unshared electrons** tend to stay in pairs. The shared electrons in the bonds and the unshared electrons repel each other, so the shape of the molecule ends up being the configuration that allows the shared and unshared electrons on the central atom to be as far apart as possible. Some of these shapes are shown in the figure, with the unshared electrons being represented by two dots.

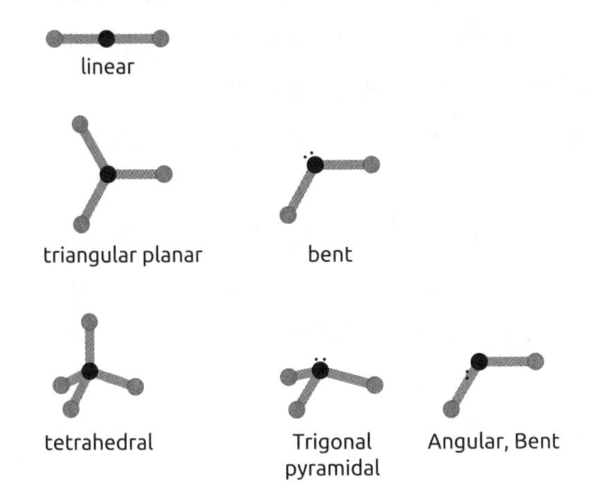

Molecules can take different shapes.

Bonding and Material Properties

Ionic compounds are held together by the strong electrostatic attraction. They therefore are hard and have high melting points. When they are dissolved in water, the solution they form conducts electricity. Covalent bonds are also very strong, but they mostly occur only between atoms in a molecule; the attraction between molecules can vary, but it is not usually as strong as the attraction between ions. Covalent compounds can, however, have a slightly positive side and a slightly negative side, due to the atoms not sharing the electrons equally. Elements more to the right of the periodic table will tend to pull the electrons toward them more, making them more negative. That will make their side of the molecule more negative, as will having extra electron pairs. These molecules are called **polar** covalent, and the attractive force between them is stronger than that between **non-polar** covalent molecules. Polar molecules tend to be hard, but not as hard as ionic compounds; they dissolve in water, but the solutions do not conduct electricity. Non-polar covalent molecules have very low melting points and do not dissolve in water.

KEY POINT!

It can be hard to visualize these compounds in three dimensions. Look for websites that show interactive models of covalent compounds so you can turn them and look at them from all sides.

Lesson Practice

TEST STRATEGY

Treat each option as a True/False question. Think about which elements can form covalent bonds with each other. Do the elements in the option match that description? True or false?

KEY POINT!

Remember that cations are positive and anions are negative. It might help you remember them if you associate the "t" in cation with the + sign.

Complete the activities below to check your understanding of the lesson content.

Skills Practice

Answer the questions based on the content covered in the lesson.

1. Which of the following pairs of elements would most likely form a covalent bond between them?
   A lithium (Li) and iodine (I)
   B sodium (Na) and sulfur (S)
   C sulfur (S) and fluorine (F)
   D aluminum (Al) and fluorine (F)

2. An atom of cesium (Cs) reacts with an atom of chlorine (Cl). Which of the following best describes what happens?
   A Cesium loses an electron, and chlorine gains an electron.
   B Chlorine loses an electron, and cesium gains an electron.
   C Cesium and chlorine share two electrons.
   D Cesium shares four electrons with two chlorine atoms.

3. Aluminum oxide has the formula Al_2O_3. This means
   A there are 3 aluminum cations for every 2 oxygen anions.
   B there are 3 aluminum anions for every 3 oxygen cations.
   C there are 2 aluminum cations for every 3 oxygen anions.
   D there are 2 aluminum anions for every 3 oxygen cations.

4. A molecule has the formula XY_3 – the element X is the central atom, and it has an unbonded pair of electrons, as well as a single bond to each of the three Ys. Which shape is the molecule?
   A linear
   B triangular planar
   C trigonal pyramidal
   D tetrahedral

5. Hydrogen and chlorine react to form a bond. They share two electrons, but chlorine pulls the electrons towards its side a bit more than hydrogen does. What type of compound do hydrogen and chlorine form?
   A ionic
   B metallic
   C non-polar covalent
   D polar covalent

Base your answers to questions 6–8 on the following passage and on the content of this lesson.

A chemistry teacher gave three unknown white powders to her students. She told them that one was sodium chloride (NaCl), one was glucose ($C_6H_{12}O_6$), and one was naphthalene, ($C_{10}H_8$), a chemical used in mothballs. She mentioned to the students that covalent compounds containing hydrogen and oxygen were often polar, while those containing only carbon and hydrogen were non-polar. The students found the melting point of each compound. They also tried dissolving each in water and tested any solutions formed for conductivity.

6. Which compound would likely have the highest melting point?
   A NaCl
   B $C_6H_{12}O_6$
   C $C_{10}H_8$
   D can't be predicted

7. Which compound(s) would be expected to dissolve in water?
   A NaCl only
   B NaCl and $C_6H_{12}O_6$
   C $C_6H_{12}O_6$ and $C_{10}H_8$
   D $C_{10}H_8$ only

8. Which compound(s) would form solutions that would conduct electricity?
   A NaCl only
   B NaCl and $C_6H_{12}O_6$
   C $C_6H_{12}O_6$ and $C_{10}H_8$
   D $C_{10}H_8$ only

See page 114 for answers and help.

KEY POINT!

Ionic compounds have high melting points and non-polar covalent compounds have low melting points.

Changes

KEY WORDS

- activation energy
- catalyst
- chemical change
- chemical equation
- chemical reaction
- combination reaction
- decomposition reaction
- endothermic reaction
- exothermic reaction
- Law of Conservation of Energy
- Law of Conservation of Matter
- physical change
- product
- reactant
- reaction rate
- synthesis reaction

KEY POINT!

Some physical changes (such as boiling) can look like chemical changes. To be certain, check to see if the identity of the substances has changed.

The only thing that seems constant in life is change: Plants grow and die; fruit ripens and then spoils. People use energy to change raw ingredients into their favorite foods. Chemists learn how to cause changes in matter and energy to obtain the results they want. They determine which changes are useful and how to speed them up, if necessary.

Chemical Reactions

If you shred a piece of paper, its form changes, but it is still paper. This is an example of a **physical change**. However, if you burn the paper, it changes into other substances, such as carbon dioxide and ash; this is an example of a **chemical change**, or **chemical reaction**. Clues that a chemical change has occurred include one or more of the following: a color change, production of a gas, production of light, and temperature change.

For example, solid sodium reacts with chlorine gas to make solid sodium chloride. We can represent this chemical reaction with a **chemical equation**:

$$2Na(s) + Cl_2(g) \rightarrow 2NaCl(s)$$

The arrow divides the reaction into "before" and "after"—the left side of the arrow shows the substances present before the reaction (called the **reactants**), and the right side of the arrow shows the substances present after the reaction (called the **products**). This type of reaction, where two or more substances combine to form one new substance, is called a **combination reaction**. It can also be called a **synthesis reaction**.

Another type of reaction is a **decomposition reaction**, where one reactant breaks apart to form two or more products, such as when carbonic acid, dissolved in water, breaks down to form carbon dioxide gas and liquid water:

$$H_2CO_3(aq) \rightarrow CO_2(g) + H_2O(l)$$

The states of matter are included with the formulas: <u>s</u>olid, <u>l</u>iquid, <u>g</u>as, <u>aq</u>ueous solution (dissolved in water). The total numbers of each type of atom must be the same on both sides of the arrow. For example, three oxygen atoms are on the left side (all in H_2CO_3), and three oxygen atoms are on the right side (two in CO_2 and one in H_2O). This is because matter is not created or destroyed during the reaction. The **Law of Conservation of Matter** states that the amount of matter remains constant.

Energy Changes

During a chemical reaction, bonds are broken in the reactant substances, and then other bonds are formed in the products. For example, the bonds in the Cl_2 molecule must be broken before the new bonds between each Cl atom and Na atom can be formed. Breaking bonds requires energy, and forming bonds releases energy; since the bonds formed are different from those broken bonds, the amount of energy being used and released can be different. If we measure the total potential energy contained in the bonds during a reaction, we see two patterns:

Endothermic Reaction

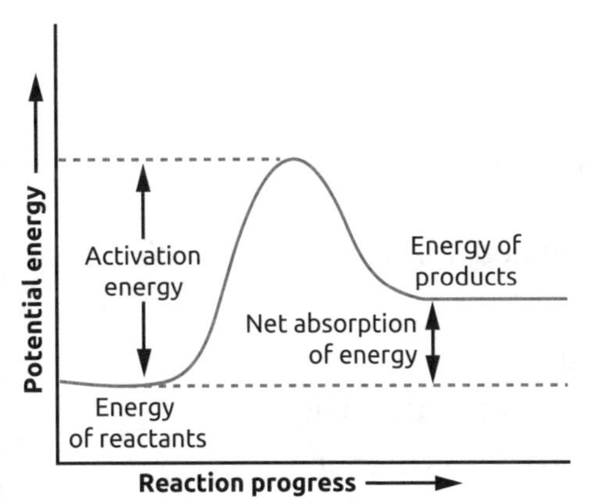

Exothermic Reaction

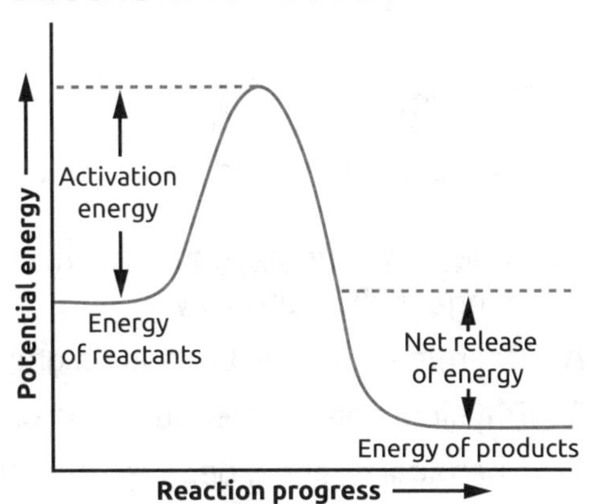

Change in energy during reactions

KEY POINT!

Remember that *exo* means "outward," *endo* means "inward," and *thermic* refers to heat. Thus, heat comes out of an exothermic reaction into its surroundings, and heat goes into an endothermic reaction from its surroundings.

For the **endothermic reaction**, you can see that the total energy of the products is higher than that of the reactants. For the **exothermic reaction**, the total energy of the products is lower than that of the reactants. However, the **Law of Conservation of Energy** states that energy cannot be created or destroyed, so the reacting substances must either be absorbing or releasing energy into their surroundings during the reaction. The energy is absorbed or released, often in the form of heat. An endothermic reaction will therefore make its surroundings cooler, while an exothermic reaction will make its surroundings warmer.

For both types of reactions, the reactants must have enough energy to react. The difference between what they have and what they need to have is called the **activation energy**. The speed at which they can get beyond the activation energy "hump" affects the **reaction rate**; this is why heating reactants often speeds up a reaction—it provides additional energy. Another way to speed up a reaction is to add a **catalyst**, which reduces the activation energy and helps the reaction proceed. Catalysts are not reactants and so are not used up in the reaction.

Lesson Practice

TEST STRATEGY

Underline key words or phrases in the question and answer choices, as shown in Question 1. The key words and phrases in the question should agree with those in the correct answer choice.

KEY POINT!

For a chemical change to occur, the identity of the substances involved must change.

Complete the activities below to check your understanding of the lesson content.

Skills Practice

Answer the questions based on the content covered in the lesson.

1. Hydrogen gas and oxygen gas, when combined, form liquid water. What type of change is this, and why?
   A Chemical change, because a gas is produced.
   B Physical change, because it is changing from gas to liquid.
   C Chemical change, because a new substance is produced.
   D Physical change, because atoms are not gained or lost.

2. Which of the following reactions is a combination reaction?
   A $2H_2O(l) \rightarrow 2H_2(g) + O_2(g)$
   B $C(s) + 2H_2(g) \rightarrow CH_4(g)$
   C $2NH_3(g) \rightarrow N_2(g) + 3H_2(g)$
   D $2NaHCO_3(s) \rightarrow H_2O(g) + CO_2(g) + Na_2CO_3(s)$

Base your answers to questions 3–5 on the following information and on the content covered in the lesson.

Plants absorb energy from the sun and use it to combine carbon dioxide gas and water to produce glucose and oxygen gas:

$CO_2(g) + 6H_2O(l) \rightarrow C_6H_{12}O_6(s) + 6O_2(g)$

They use the glucose molecules to store the energy for later use.

3. How many molecules of carbon dioxide should be on the reactant side to have an equal number of carbon atoms on each side?
   A 1
   B 3
   C 6
   D 12

4. Which law requires there to be the same number of each type of atom on each side?

   ____

5. Would you expect this reaction to be exothermic or endothermic? Why?

   ____

   ____

   ____

Base your answers to questions 6–8 on the following information and on the content covered in the lesson.

A group of chemists are working together to make sports wraps that can become cold on their own, providing relief from injuries. They plan to put two small bags of chemicals inside a large bag of gel, because when the inner bags are squeezed, the bags break and the chemicals react, cooling the gel. They measured the energy changes for the reactions of two different sets of chemicals they are considering:

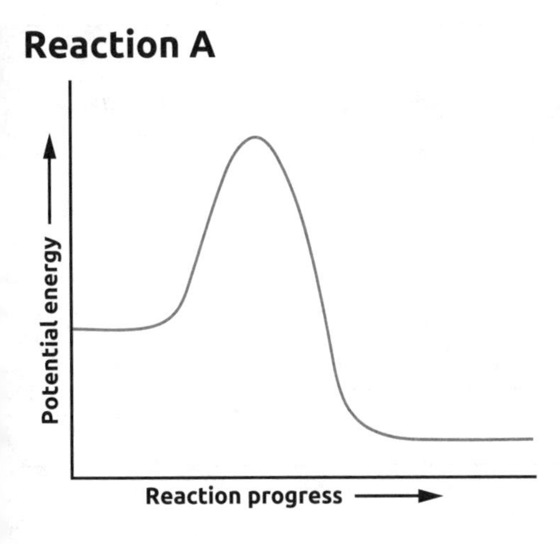

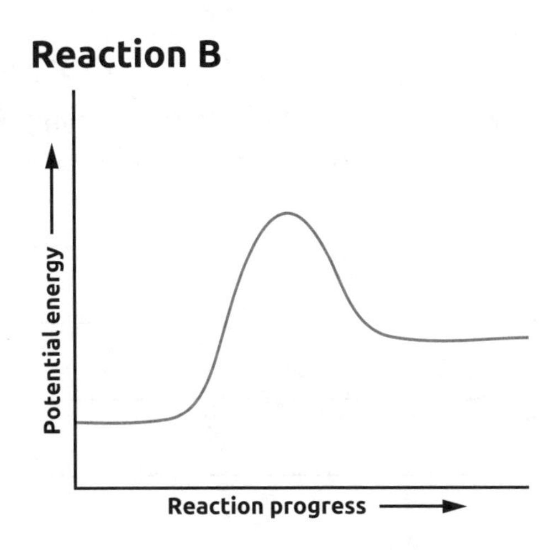

6. Is an endothermic or exothermic reaction needed for their new product? Why?

____

____

____

____

KEY POINT!

In an exothermic reaction, energy is released; in an endothermic reaction, energy is absorbed.

Lesson Practice

7. Based on the graphs, would Reaction A or Reaction B work best for their product? How do you know?

____

____

____

____

8. The chemists would like the cooling reaction to happen very quickly. What can they do to speed it up? Why would that work?

____

____

____

____

See page 114 for answers and help.

All things are made of matter, and there are many types of matter, as well as many different ways to combine matter. Knowing how to classify matter can help a chemist predict what properties an unknown material will have.

Phases and Phase Changes

All matter can exist as a **solid**, **liquid**, or a **gas**, but the temperatures at which these states exist change from substance to substance. For example, at 30°C, glucose is a solid, water is a liquid, and carbon dioxide is a gas. Warming a substance requires energy, as does going from solid to liquid or liquid to gas.

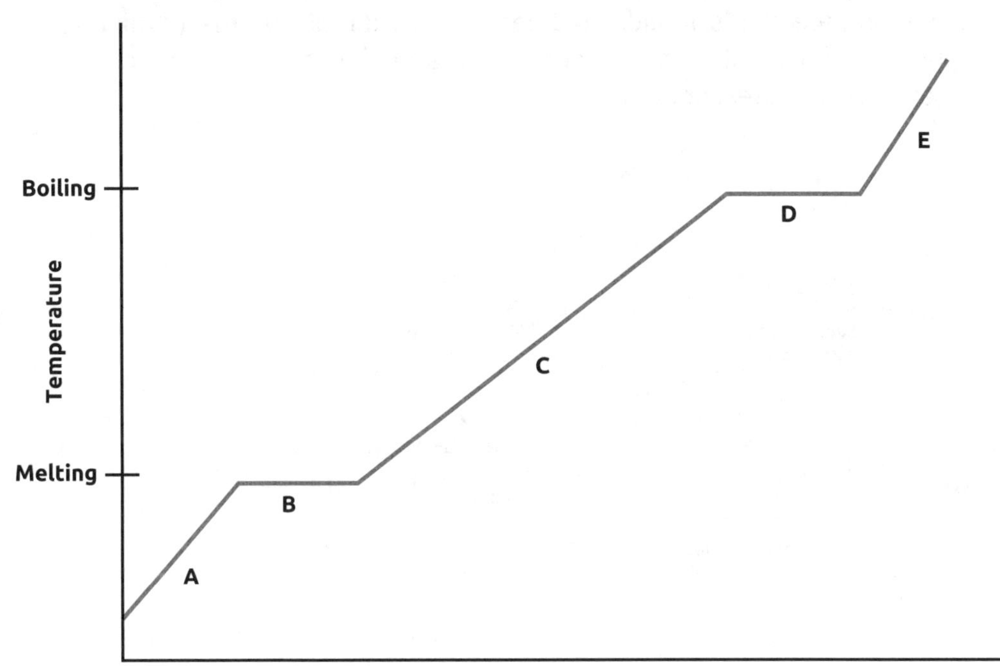

General heating curve

The graph shows that, as heat is added to a solid (section A), the temperature of that solid increases until it gets to the **melting point**. While the substance is melting (B), it continues to absorb heat, but instead of raising the temperature, the heat energy is used to disrupt the orderly structure of the solid. Once all of the particles in the solid can move around freely, the substance is a liquid, and the temperature begins to rise again (C). When it reaches the **boiling point** (D), the heat energy is used to overcome any attraction between the particles, so they can move apart. Once the attraction is overcome in all of the particles, the substance is a gas, and the temperature begins to rise again (E). If the substance were water, B would occur at 0°C, and D would occur at 100°C. If, instead, the substance started out as a gas and heat was being removed, it would be a cooling curve, with every event happening in reverse. The gas would **condense** into a liquid at the same temperature at which it boils, and then it would **freeze** into a solid at the same temperature at which it melts.

KEY WORDS

- acid
- base
- boiling point
- colloid
- condense
- freeze
- gas
- heterogeneous
- homogeneous
- liquid
- melting point
- mixture
- pH scale
- solid
- solute
- solution
- solvent
- suspension

KEY POINT!

At the melting/freezing point, both solid and liquid will be present; at the boiling/condensing point, both liquid and gas will be present.

KEY POINT!

An easy way to tell the difference between compounds and mixtures: compounds have formulas, while mixtures have a list of ingredients.

Classification of Matter

Matter can be classified based on its composition. A compound is different from a mixture in that a compound is made of elements that have reacted chemically to form a new substance; a **mixture**, on the other hand, is made of substances that have been physically blended but retain their identities. Mixtures have two or more substances in them and can have different properties, depending on what those substances are. A **homogeneous** mixture, also known as a **solution**, looks the same throughout, while the components of **heterogeneous** mixtures can be seen. The parts of a heterogeneous mixture can stay mixed, making it a **colloid**, or they can separate, making the mixture a **suspension**. The examples in the diagram are solids, liquids, and gases. For example, bronze (a mixture of copper and tin) is solid, salt water is liquid, and air is a gas—they are all homogeneous mixtures, or solutions.

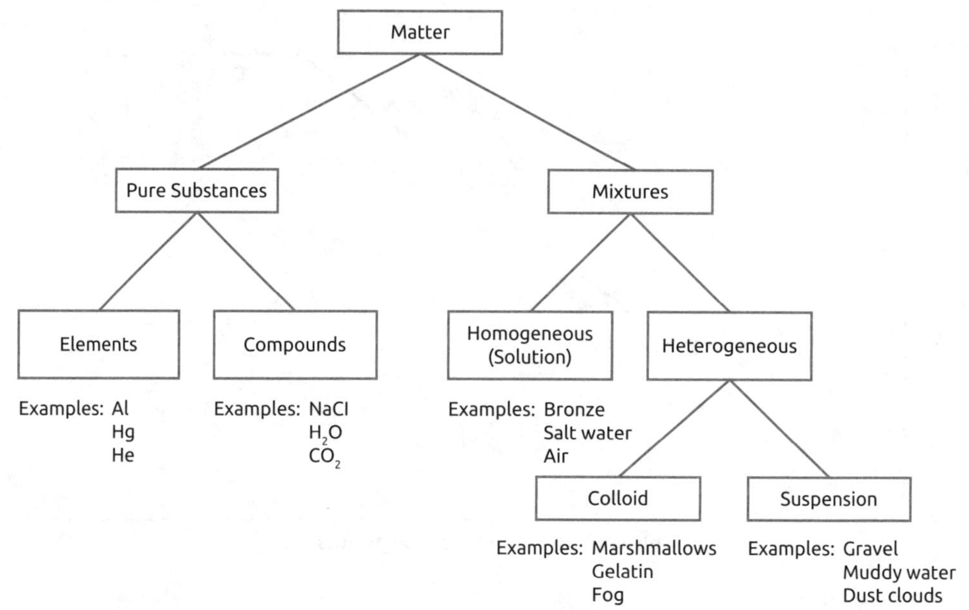

Classification of matter

Properties of Solutions

In a solution, the substance present in the largest quantity is considered the **solvent**; all the other substances dissolved in the solvent are solutes. For example, in a solution of salt water, the water is the solvent and the salt is the **solute**. The amount of solute that can be dissolved depends on the solute and solvent, as well as the temperature. At 20°C, 100g of water can dissolve about 36g salt, while at 100°C, 100g of water can dissolve about 39g salt. At the same temperatures, 179g sugar and 487g sugar can be dissolved.

If the solute is an **acid** or a **base**, the solution is referred to as acidic or basic. A pH measurement indicates how acidic or basic the solution is—the **pH scale** runs from 0 (most acidic) to 14 (most basic) with a pH of 7 being neutral. Pure water is neither acidic nor basic; it has a pH of 7.

Complete the activities below to check your understanding of the lesson content.

Skills Practice

Answer the questions based on the content covered in the lesson.

1. A chemist has a solution that has a pH of 4 and would like to neutralize it. Which would be the best way to neutralize it?
   A Add an equal amount of water.
   B Add an equal amount of a pH 9 solution.
   C Add a small amount of acid.
   D Add a small amount of water.

2. Ethanol has a melting point of –114°C and a boiling point of 79°C. At which of these temperatures would ethanol be a liquid?
   A –200°C
   B –120°C
   C –85°C
   D 90°C

3. Which of the following is a compound?
   A ammonium nitrate (NH_4NO_3)
   B carbonated water (CO_2 and H_2O)
   C brass (Cu and Zn)
   D salt water (NaCl and H_2O)

4. A cook made salad dressing by mixing oil and vinegar. When she stops mixing, it separates into two layers. Which classification of matter is the salad dressing?
   A compound
   B solution
   C colloid
   D suspension

5. A candy maker needs to dissolve a lot of sugar in some water. She added about 300g sugar to about 100g water at 20°C and stirred, but not all of it dissolved. Which would be the best way to get the rest of the sugar dissolved?
   A Stir the mixture longer.
   B Heat the mixture.
   C Add more sugar.
   D Wait until some water evaporates.

TEST STRATEGY

Note the meaning of any terms you know. Then, try to determine the meaning of unfamiliar terms by looking at prefixes and suffixes, or by seeing if they are similar to words you do know. If you know what *neutral* means, can you guess what *neutralize* means?

KEY POINT!

A substance will be a liquid at temperatures above the freezing/melting point and below the boiling/condensing point.

Lesson Practice

KEY POINT!

If you can see where parts of a mixture are not the same throughout, it is heterogeneous.

Base your answers to questions 6–8 on the following graph and on the information in the lesson.

A chemist has a sample of hot gas and measures the amount of heat being released as it cools.

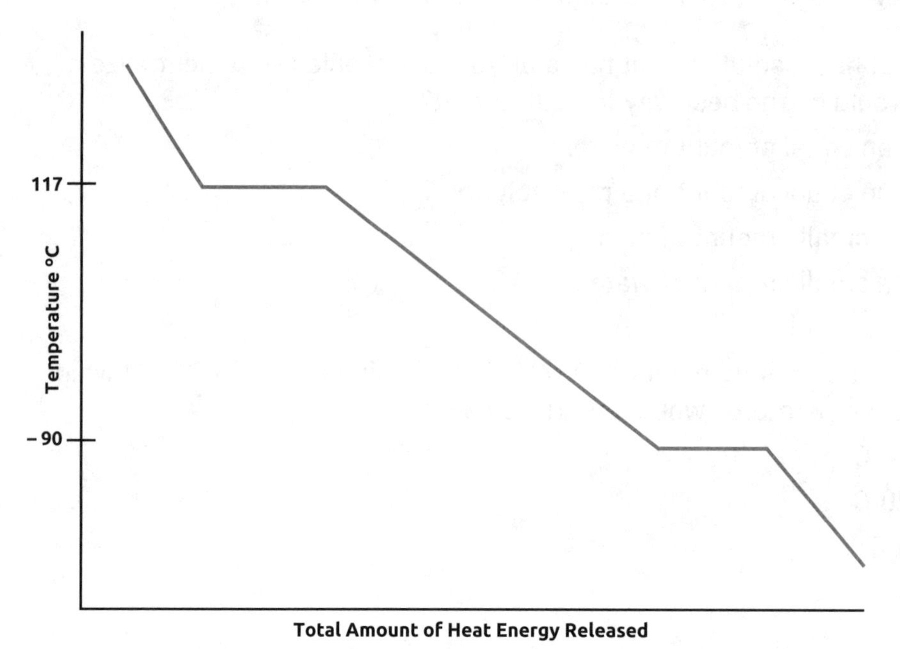

Cooling curve of gas

6. What is occurring at –90°C?
   - **A** The substance is boiling.
   - **B** The substance is freezing.
   - **C** The substance is cooling down.
   - **D** The substance is heating up.

7. Which state(s) of matter are found at 117°C?
   - **A** solid and liquid
   - **B** liquid and gas
   - **C** liquid only
   - **D** gas only

8. What state of matter will the substance be at –65°C? How do you know this?

___

___

___

___

See page 114 for answers and help.

Energy is everywhere. Sometimes it is in a form we can use, and sometimes we must convert it to a useful form. Knowing what forms of energy are available and to what forms they can be converted helps us make sure we have enough energy. We can also look for sources of energy that are safer and better for the environment, or sources that cost less.

KEY WORDS

- batteries
- hydroelectric
- kinetic energy
- light energy
- nuclear power
- potential energy
- solar
- wind power

Types of Energy

Energy can be classified as either kinetic energy or potential energy. **Kinetic energy** involves motion, while **potential energy** is stored energy. The kinetic energy contained in a moving object is proportional to both its mass and its velocity:

$$KE = \frac{1}{2}mv^2$$

This means that if two cars are traveling at the same velocity, the one with more mass will have more kinetic energy. Likewise, if the cars have the same mass, the one that is traveling faster will have more kinetic energy.

A falling object increases in velocity as it falls, so the longer it falls, the more kinetic energy it has. Therefore, the higher an object is, the more potential it has to gain kinetic energy by falling. This potential to gain energy is known as gravitational potential energy. It is proportional to three things: the mass of the object, the acceleration of gravity, and the height of the object:

$$PE = mgh$$

On Earth, the acceleration of gravity is 9.8 m/s^2. On the moon, it is 1.6 m/s^2, and on Mars, it is 3.7 m/s^2. This means that if you compare the potential energy of objects that are the same mass and the same distance above each planet, the one on Earth will have the most potential energy, and the one on the moon will have the least amount of potential energy.

Energy Conversion

Potential energy (PE) can be converted to kinetic energy (KE); likewise, kinetic energy can be converted to potential energy. As an object falls, its gravitational PE is converted to KE. Roller coasters operate on this principle. The car in the following image has minimal KE but maximum PE. As the car starts down a slope, the PE converts to KE, and the car moves faster. Once the car starts climbing again, the KE converts to PE. The car moves more slowly as it climbs higher, so the next hill must be low enough that the car reaches the top before its velocity decreases to zero.

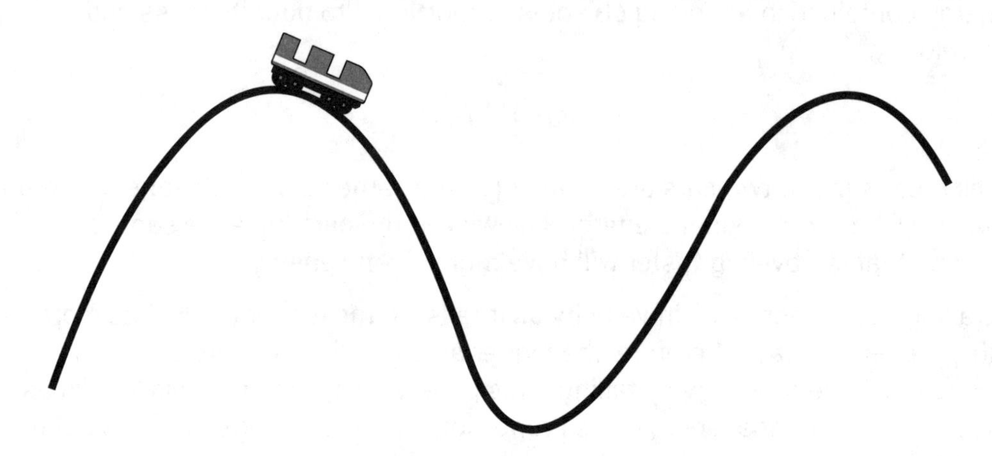

PE and KE change as the car changes height.

Hydroelectric plants convert the potential energy from falling water to kinetic energy in different forms. The falling water turns a turbine. This moves magnets and conductors in the generator, which generates an electric current. Since electricity involves the movement of electrons, it is a form of kinetic energy. **Wind power** and **nuclear power** work in similar ways. In wind power, wind turns the turbine, which powers the generator. In nuclear power plants, the heat from nuclear decay heats water to steam. The steam then moves turbines and powers the generator. Other power plants use fossil fuels, such as coal, oil, and natural gas, to produce the heat to generate electricity.

Light energy is a form of kinetic energy. Visible light and infrared light (heat) are produced when fuel is burned. They can also be produced by converting the kinetic energy from electricity in the wire through which the electricity runs. This causes the filament in light bulbs or the "burner" on electric stoves to glow.

Light energy can also be used to generate electricity with **solar** panels. The light energy from the sun excites electrons in the material of the solar panel. The panel is made in such a way that the electrons can be "herded" in one direction, producing a current. Electrical energy can be produced by chemical reactions, and chemical reactions can be caused by electricity. **Batteries** have the potential energy of chemical reactions. They will release electrical energy when they are hooked up to a circuit.

Complete the activities below to check your understanding of the lesson content.

Skills Practice

Answer the questions based on the content covered in the lesson.

1. A roller coaster is at the bottom of a hill, about to start up. It has a mass of 100 kg and a velocity of 16 m/s. How high can the car climb up the next hill before it stops? [g = 9.8 m/s^2]
   A 1 meter
   B 13 meters
   C 156.8 meters
   D 1600 meters

2. A waterfall is 100 m high. How much kinetic energy will each 1 kg of water provide to a turbine at the bottom?
   A 50 kg m^2/s^2
   B 100 kg m^2/s^2
   C 980 kg m^2/s^2
   D 2000 kg m^2/s^2

3. Hydroelectric power and wind power are both considered renewable energy sources. In what other way are they similar?
   A They both require changes in height.
   B They both rely on chemical reactions.
   C They both convert potential energy to kinetic energy.
   D They both convert one form of kinetic energy to another form of kinetic energy.

4. Cars burn gasoline, and the heat of the expanding gas moves the engine, which moves the wheels. What kind of energy conversion happens when the gasoline is burned?
   A potential to kinetic
   B potential to electrical
   C kinetic to potential
   D kinetic to electrical

TEST STRATEGY

Rewrite the question in your own words to make sure you understand it.

KEY POINT!

Remember that kinetic energy is the energy of motion. If an object or substance is moving, it has kinetic energy; if not, then it doesn't.

Lesson Practice

5. What kind of energy conversion happens in a solar cell phone charger?
   A potential to kinetic to electrical energy
   B kinetic to electrical to potential energy
   C potential to electrical to kinetic energy
   D kinetic to potential to electrical energy

6. Hydroelectric plants at dams often pump water back uphill to behind the dam during the night, when there is not much demand for electricity. What is the best explanation for why they do this?
   A It helps balance the difference in heat energy between the two spots.
   B It helps increase the kinetic energy of the water remaining below the dam.
   C They are increasing the potential energy of the water by using excess kinetic energy.
   D They can use the kinetic energy being lost by the water as it is pumped uphill to increase the amount of electrical energy available.

See page 115 for answers and help.

We use energy all the time, in many forms; we turn on the stove to heat a pot of water for tea, we put food in the microwave, we use solar energy to charge our cell phone batteries. Energy is all around us and cannot be destroyed, no matter how much we use it; it can, however, be converted from one form to another. Many of the devices we use each day work by converting one form of energy into a more useful form of energy.

KEY WORDS

- circuit
- conduction
- convection
- current
- electricity
- electromagnetic spectrum
- frequency
- radiation
- wavelength

Transferring Energy

Energy in the form of heat is transferred three ways: convection, conduction, and radiation. **Convection** occurs when heat is transferred through fluid moving away from a heat source, spreading out the heat energy. Convection sounds complicated but is relatively easy and can be seen in everyday life—for example, in a pot of water being heated up, the water in the bottom of the pot gets hot and then rises. Remember that heat always rises because the molecules are farther apart and thus less dense. The rising hot water then cools as it gives some of its heat to the cooler water at the top, which also releases some of its heat to the air above it. The cooler water then starts to sink, starting the process over again.

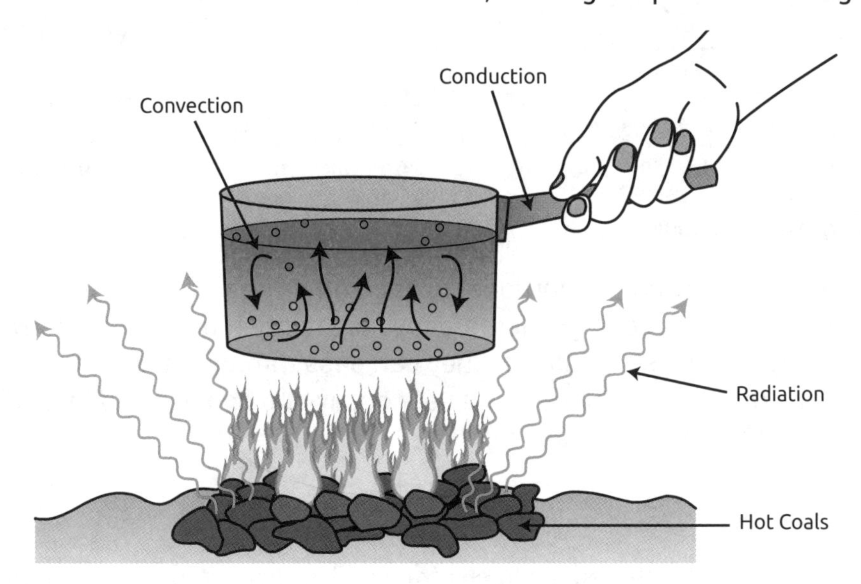

Convection, conduction, and radiation in a pot of water.

Conduction occurs when the vibrating molecules of a hot surface touch another cooler surface. This causes the molecules of the second surface to start vibrating faster. Remember that when molecules are heated, they vibrate faster. When one molecule "runs into" a slower molecule, it transfers energy to that molecule. The second molecule is now vibrating faster than before the collision, causing it to heat up. In the image, conduction is happening at the handle of the pan. The heat energy is moving—bumping molecules transfer energy through the pot material from the bottom, where it was heated when it was touching the hot coals, up through the handle.

Radiation is energy transfer that does not utilize matter. In the image, the heat energy of the burning coals can be felt by the hand, but it is not carried by air currents or through the metal of the pot; the thermal energy is transferred as waves. Since radiation does not need to pass through matter, solar radiation is able to reach Earth through miles of airless space, to be absorbed by Earth and its atmosphere.

Electromagnetic Spectrum

The **electromagnetic spectrum** is the range of waves that carry different forms of electromagnetic radiation. Infrared radiation carries heat, and the visible light portion of the spectrum is the only part we can see. **Wavelength** and **frequency** are inversely proportional to each other, as described by this equation:

$$\upsilon = \lambda f$$

Here, υ is the velocity of the wave (the speed of light, which is constant), λ is the wavelength, and f is the frequency. The longest waves have the lowest frequency.

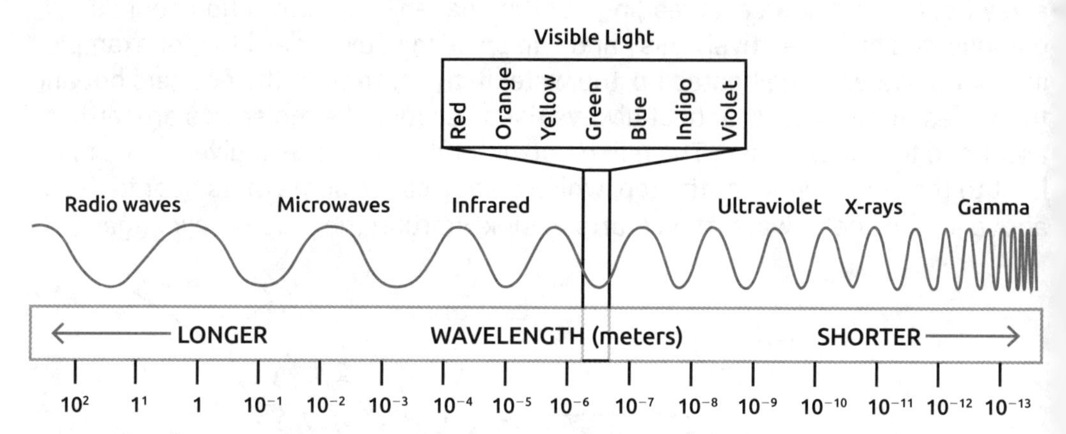

Electromagnetic spectrum

Energy and frequency are proportional to each other, so the higher frequency wavelengths have more energy. Gamma rays are dangerous because they carry a lot of energy and are so small that they can pass through cells and damage DNA. Gamma rays are produced by nuclear decay and can be used in medical procedures such as PET scans.

Electric Currents

Energy can also be transferred using **electricity**. Electricity carries the electrical potential energy from the storage site, such as a battery, to the device where the energy is used, such as a light bulb. The device converts the electrical energy into another form of energy. For example, the light bulb converts the electrical energy into heat and light. Electricity is the flow of an electric charge. It is often represented as the flow of electrons through a circuit. A **circuit** is a closed loop. The simplest circuit includes just a battery, a device, and two wires forming a circle.

Electrons flow from the battery to the light bulb, carrying electrical energy. The energy is used by the light bulb to produce light, and the electrons then flow back into the battery. This flow of electrons is called a **current**.

Complete the activities below to check your understanding of the lesson content.

Skills Practice

Answer the questions based on the content covered in the lesson.

1. A camper constructed a solar shower by hanging a bag of water from a tree so that the sun was shining on the lower part of the bag. After 4 hours, all the water inside the bag, including in the upper part, was warm. How did this transfer of energy occur?

   A radiation from the sun, convection through the bag, conduction through the water

   B convection from the sun, conduction through the bag, radiation through the water

   C radiation from the sun, conduction through the bag, convection through the water

   D convection from the sun, radiation through the bag, conduction through the water

2. How is energy transferred through convection?

   A energy waves moving through space

   B energy moving from one molecule to another after a collision

   C energy moving with matter as it moves upward due to a change in density

   D energy waves being converted from one type of energy to another

3. Ultraviolet waves have less energy than _______.

   A x-rays

   B visible light

   C microwaves

   D radio waves

4. Sunscreens work by blocking ultraviolet radiation. You can see through the sunscreen, and you still get hot while wearing it in the sun. People who wear sunscreen when outdoors reduce their risk of skin cancer. Which is the best explanation for this?

   A Sunscreens absorb all radiation, preventing damage to the skin.

   B Sunscreens absorb ultraviolet, visible, and infrared radiation, preventing damage to the skin.

   C Ultraviolet radiation has a high enough wavelength to cause damage, while infrared and visible do not.

   D Ultraviolet radiation has a high enough frequency to cause damage, while infrared and visible do not.

TEST STRATEGY

Cover the answers to the question. Think of what you would write if the answers weren't there, and then uncover the answers. Choose the one that best matches your own answer.

Lesson Practice

Select the best option.

5. A hair dryer converts (electric, heat) energy to (electric, heat) energy.
6. A switch can be used to shut off electric current by making the circuit (complete, incomplete).
7. An electric current is caused by the flow of (atoms, electrons) through a circuit.
8. A battery (produces, stores) electrical energy.

See page 115 for answers and help.

We utilize the laws of motion every day without even realizing it. If a small push is ineffective at making a shopping cart move fast enough, we give it a stronger push. When we are driving a car at a high speed, we know it is necessary to have extra space and time to slow down and stop than when we are driving more slowly. We know it is more difficult to displace a heavy object than it is to move a light one. We discover these things by experience, without calculations, but if we want to accurately predict motion, we need equations.

KEY WORDS

- acceleration
- distance
- force
- mass
- momentum
- net force
- velocity

Equations of Motion

The calculations needed to understand motion can seem confusing at first glance. The best approach is to write down each variable given in the problem, and then choose the equation that uses those variables. These tables offer a summary of the main variables and equations:

Variables	
d	distance
t	time
v	velocity
a	acceleration
m	mass
F	force
ΣF	net force
p	momentum

Equations	
Velocity, distance, and time	$v = \frac{d}{t}$
Acceleration, velocity, and time	$a = \frac{v_{end} - v_{begin}}{t}$
Force, mass, and acceleration	$F = ma$
Multiple forces acting	$\Sigma F = F_1 + F_2 + \ldots$
Momentum, mass, and velocity	$p = mv$

Distance is a measurement of how far something has traveled. **Velocity** is how fast something is traveling in a certain direction, as in the distance traveled north in a certain amount of time. Velocity is similar to speed, but it includes direction. The direction becomes important when working with forces.

Example: John drives a distance of 2,500 meters to the pharmacy. If it takes him 10 minutes, what was John's average velocity in m/s (meters per second)?

> **ANSWER:** The question mentions distance, velocity, and time, so we use the first equation. We know that d = 2500 meters and t = (10 minutes)(60 seconds/minute) = 600 seconds, so we can plug those numbers into the equation for speed to get $v = \frac{2500\text{m}}{600\text{s}} = 4.17\ \text{m/s}$ in the direction of the pharmacy.

Acceleration is how fast the velocity of something is changing. It can also be described as the amount the velocity has changed in a certain amount of time.

Example: On his way home, John decides to increase his velocity, and it takes him 30 seconds to increase from 4.17 m/s to 8 m/s. What is John's acceleration?

Answer: This example has two velocities, time, and acceleration, so we use the second equation. If $v_{end} = 8\text{m/s}$ and $v_{begin} = 4.17\text{m/s}$ and t=30 seconds, then

$$a = \frac{8\text{m/s} - 4.17\text{m/s}}{30\text{s}} = 0.128\frac{\text{m/s}}{\text{s}} = 0.128\frac{\text{m}}{\text{s}^2}$$

Mass is the quantity of matter contained in an object, while **force** is a push or a pull on an object. If you attempt to push or pull an object to displace it, it requires a particular amount of force; if you try to move a heavier object (one with more mass), it requires more force; if you attempt to make the object move faster, it also involves more force. Therefore, force is related to both mass and acceleration.

Example: Kianna is pushing her brother on a swing. Her brother accelerates 0.5 m/s^2 and has a mass of 30 kilograms. With what force is Kianna pushing?

ANSWER: Use the third equation. Force is usually measured in Newtons (N), which are $1\frac{\text{kg m}}{\text{s}^2}$. If $a = 0.5\frac{\text{m}}{\text{s}^2}$ and $m = 30$ kg, then $F = 0.5\frac{\text{m}}{\text{s}^2} \times 30\text{kg} = 15\frac{\text{kg m}}{\text{s}^2} = 15\text{N}$.

If two forces are acting on an object, the **net force** is the sum of the two forces. If they are acting in opposite directions, they will have opposite signs. For example, consider two children, Matt and Kelly, both pushing on a box from opposite sides. However, Matt pushes with a force of 3 N and Kelly pushes with 5 N, so the net force is 2 N in the direction that Kelly is pushing. The force of friction will act as a force pushing in the opposite direction of the motion.

Example: Aimee is pulling a sled along the road toward a hill. She is pulling with a force of 6 N, and the ground friction is acting as force of 1 N. What is the net force?

ANSWER: $\Sigma F = 6\text{N} - 1\text{N} = 5\text{N}$

Momentum describes the amount of movement energy an object has. An object that is very heavy or moving very fast has more momentum than a smaller or slower object, and it requires a larger force to stop or start movement.

Example: A truck has a mass of 2,000 kilograms and is traveling at 70 kilometers/hour (19.4 m/s). What is its momentum?

ANSWER: If m = 2000 kg and v = 19.4 m/s, then p = 2000 kg x 19.4 m/s = 3880 kg m/s.

The truck would be much harder to stop than a small car traveling at the same velocity.

Complete the activities below to check your understanding of the lesson content.

Skills Practice

Answer the questions based on the content covered in the lesson.

1. Jessica is pulling on a heavy cart with a force of 40 N, while Julia and Allen help by pushing it in the same direction with a force of 30 N each. The force of friction acting on the cart is 15 N. What is the net force moving the cart forward?

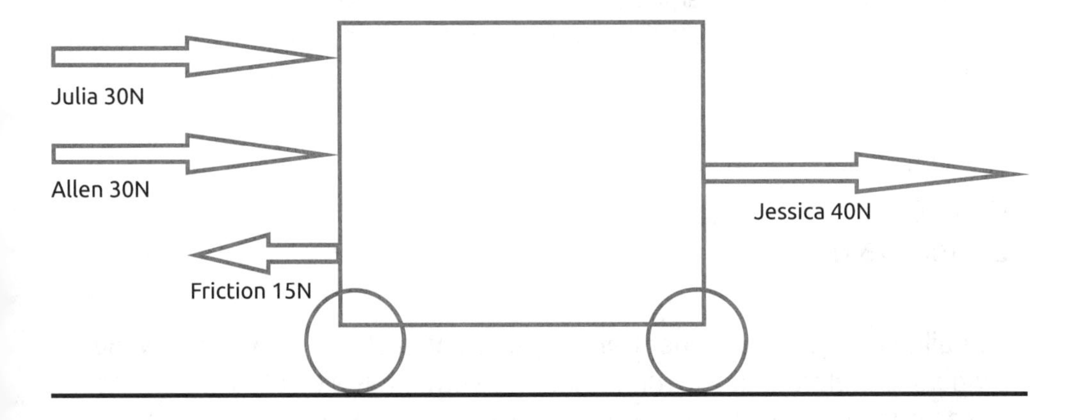

A 25 N
B 45 N
C 85 N
D 115 N

2. Usain Bolt ran the 100-meter dash in 9.63 seconds in the 2012 Olympics. What was his average velocity?

A 9.63 m/s
B 10.4 m/s
C 80.37 m/s
D 96.3 m/s

KEY POINT!

Look at the information given. What variables does it include? Use an equation that includes those variables and plug in the known values to solve for the missing one.

TEST STRATEGY

Make a sketch that represents the actions in the problem; include numbers and arrows to indicate motion/forces and direction.

Lesson Practice

3. An amateur runner is trying to improve her time in the 100-meter dash. She found that she can reach a velocity of 8.5 m/s in the first 2 seconds of the race. What is her acceleration?

   A 4.25 m/s^2

   B 6.5 m/s^2

   C 8.5 m/s^2

   D 17.0 m/s^2

4. The runner's coach tells her that she needs to push harder against the starting blocks as she begins the race in order to get an immediate acceleration of 5.5 m/s^2. If the mass of the runner is 65 kg, what force will she need to use?

   A 0.18 N

   B 59.5 N

   C 357.5 N

   D 1966.25 N

5. To build strength, the runner trains by pushing a sled with weights while trying to accelerate. If the runner pushes with a force of 115 N while the coach pushes back from the other side with a force of 25 N, and the force of friction acting on the sled is 27 N, what is the net force on the sled?

   A 52 N

   B 63 N

   C 113 N

   D 167 N

6. The runner gets faster and reaches a velocity of 9.8 m/s. Her mass is 65 kg. What is her momentum?

   A 6.63 kg m/s

   B 55.2 kg m/s

   C 74.8 kg m/s

   D 637 kg m/s

See page 115 for answers and help.

After a long day of studying, you might feel hungry and exhausted and think you've accomplished a lot of work, but from a scientific point of view, you have just sat around reading all day. In fact, physics would determine that you have done little to no actual work. However, if you had spent the day moving furniture around your living room and carrying loads of laundry up and down the stairs, then, according to science, you have done work. According to the laws of physics, in order to accomplish work, you need to use a force to move something: a box, a couch, a basket of laundry.

KEY WORDS

- fluid friction
- friction
- fulcrum
- gravitational force
- inclined plane
- lever
- magnetic force
- pulley
- rolling friction
- sliding friction
- static friction

Types of Forces

A force is a push or pull on an object. We can observe various types of forces; for instance, all objects have gravity, a force of attraction. The greater the mass and the closer the objects are, the greater the **gravitational force** between them. Earth's gravity is what causes you to fall down when you stumble, and it also keeps the moon revolving around Earth. **Magnetic force** is the force of attraction between magnets; it is what causes them to "stick" when brought together.

Friction is a force that pushes against, or resists, motion. Different surfaces have different amounts of friction—for instance, it is easier for you to slide across a wood floor in socks than it is to slide across the same floor in rubber-soled sneakers. There are several types of friction, including static friction, sliding friction, rolling friction, and fluid friction.

The word "static" means unchanging: **static friction** exists between two objects that are touching but not moving. If you have ever tried to move a heavy box from an unmoving position, you have encountered static friction; it is the force that makes that initial movement so difficult.

As the name implies, **sliding friction** is the force that is present when one object slides over another object. Take the box that you started moving in the last example—once you overcome static friction, you encounter sliding friction as you slide the box across the floor. If you were to put wheels on the box, it would be easier to move; **rolling friction**, the friction that exists between a round object and another surface, is weaker than sliding friction.

A fourth type of friction, **fluid friction**, is the force that opposes movement through a fluid, such as water. For instance, you must work against fluid friction when you swim, and a plane encounters fluid friction as it flies through the air, since gases and liquids are both considered fluids.

Work

Work is defined as the amount of energy required to move an object. Consider that box once again. When you push on it and it begins to slide across the floor, work is being done; a force (your pushing) causes a change in motion (the box moving from one position to another). If you push on that box and it does not change its position, no work has been done; the force must cause the object to move in the same direction as the applied force. Work depends on force and distance; the more force exerted or the greater the distance traveled, the more energy required (the more work done).

Simple Machines

Simple machines are devices that make moving something with force (doing work) easier; they accomplish this by changing the direction or the size of the force applied.

Inclined Plane

Consider lifting a heavy or awkward object into the back of a truck; it is much easier to slide the object up a ramp than it is to pick it up and lift it over your head. You will have to slide the object over a longer distance than going straight up, but less force will be necessary. The same amount of work is being done, but over a longer distance, so less force needs to be applied. A wheelchair ramp and a slide are examples of **inclined planes**.

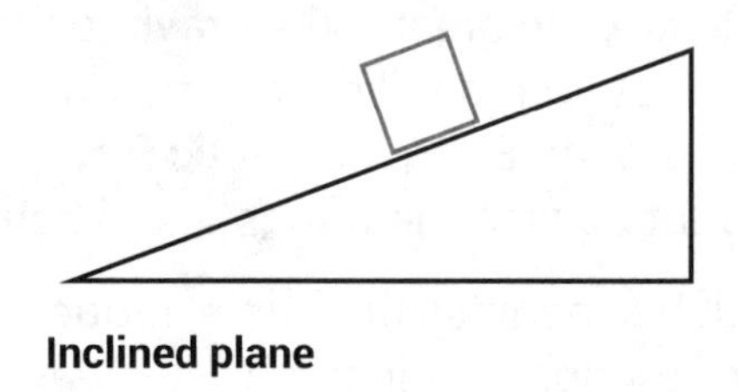

Inclined plane

Lever

A **lever** is a simple machine that consists of a bar with a fixed support, called a **fulcrum**, that allows the bar to pivot.

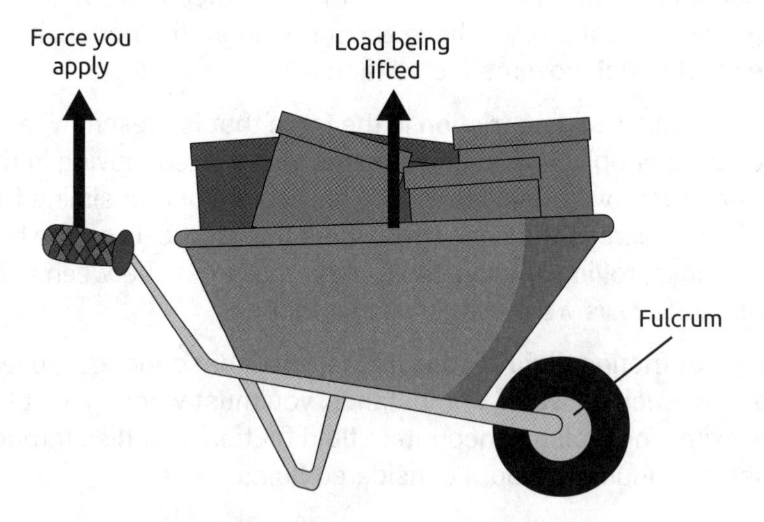

The handle of a wheelbarrow is an example of a lever.

Pulley

A **pulley** consists of two sections: a rope that can be pulled and a grooved wheel into which the rope settles. If there is a single pulley or the pulley is stationary, it changes only the direction of the force; if there are multiple pulleys or the pulley is moveable, the size of the force required can be decreased.

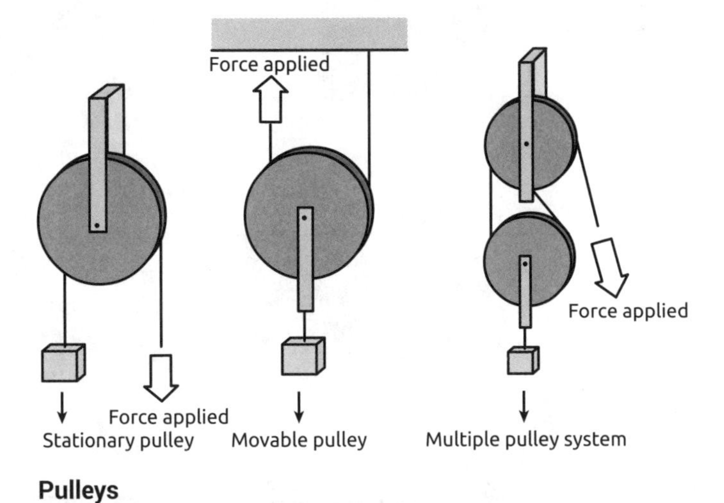

Pulleys

Lesson Practice

UNIT 3 / LESSON 8

Complete the activities below to check your understanding of the lesson content.

Skills Practice

Answer the questions based on the content covered in the lesson.

1. Which type of friction is shown in the image below?

A fluid friction
B rolling friction
C sliding friction
D static friction

Lesson Practice

TEST STRATEGY

Treat each answer option as a true/false question. Jot down why each is true or false.

Use the following image to answer questions 2–5.

2. What type of friction exists between the box and the floor in part 1 of the image?
   A fluid friction
   B rolling friction
   C sliding friction
   D static friction

3. In which picture is the person doing work?
   A part 1
   B part 2
   C part 3
   D part 4

4. What type of simple machine is used to help move the box?
   - **A** fulcrum
   - **B** lever
   - **C** pulley
   - **D** inclined plane

5. What makes the work easier to accomplish?
   - **A** The box is being moved a greater distance, so less force is required.
   - **B** The box is being moved a greater distance, so less work is being done.
   - **C** The box is being moved a shorter distance, so less force is required.
   - **D** The box is being moved a shorter distance, so less work is required.

6. Which is an example of gravitational force?
   - **A** Two magnets come together.
   - **B** Two magnets are forced apart.
   - **C** A student drops a magnet on the floor.
   - **D** A person pushes a magnet across a table.

7. A lever pivots about a ________.
   - **A** friction
   - **B** fulcrum
   - **C** midpoint
   - **D** pulley

8. Under what conditions does a pulley reduce the amount of force needed to move an object?
   - **A** when the pulley is stable
   - **B.** when the pulley is moveable
   - **C** when the rope is pulled upward
   - **D** when the rope is pulled downward

See page 115 for answers and help.

Answer the questions based on the content covered in this unit.

Use the following passage to answers questions 1–3.

Recipe for Pancakes

In a mixing bowl, combine 3 cups flour, 3 tablespoons sugar, 5 teaspoons baking soda, and ½ teaspoon salt. Mix these well, and then add 3½ cups buttermilk, 3 eggs, and ⅓ cup melted butter. Stir until just blended. Pour about ¼ cup of the batter onto a hot griddle for each pancake. Use a spatula to peek at the bottoms of the pancakes. Turn them when the bottoms are light brown. When the other side is light brown, remove them from the griddle, then serve.

You can alter this recipe to suit your diet or the availability of ingredients. For example, if you are on a low-sodium diet, you may want to substitute potassium chloride for the sodium chloride (salt). If you don't have any buttermilk, you can use regular milk, but you will need to add some vinegar or lemon juice to lower the pH of the milk. If you don't feel like melting butter, you can use vegetable oil, which is already a liquid. Enjoy your pancakes!

1. The recipe says that you can substitute potassium chloride for sodium chloride. Which best explains why potassium chloride is similar to sodium chloride?
   A Potassium atoms have the same number of protons as sodium atoms.
   B Potassium and sodium atoms are the same size.
   C Potassium and sodium are different forms of the same substance.
   D Potassium is in the same group as sodium.

2. The recipe says that milk can be used instead of buttermilk, but you must first lower the pH of the milk. What does this suggest?
   A Buttermilk is more acidic than milk.
   B Buttermilk is less acidic than milk.
   C Buttermilk is neutral, and milk is acidic.
   D Buttermilk is basic, and milk is neutral.

3. According to the recipe, you need to melt butter to use it, but you don't need to melt oil, as it is already a liquid. What does this tell you?
   A The melting temperature of butter is above room temperature.
   B The melting temperature of butter is below room temperature.
   C The melting temperatures of both butter and oil are above room temperature.
   D The melting temperatures of both butter and oil are below room temperature.

4. Molybdenum (Mo) is element 42 on the periodic table. What set of properties would you expect molybdenum to have?
   A shiny, brittle, conducts heat and electricity well
   B dull, brittle, does not conduct heat and electricity well
   C shiny, malleable, conducts heat and electricity well
   D dull, malleable, does not conduct heat and electricity well

5. Which particle is the smallest?
   A atom
   B electron
   C neutron
   D proton

6. Which pair of elements is most likely to form an ionic bond?
   A C and O
   B Li and Br
   C N and I
   D S and F

7. Which best describes a covalent bond?
   - **A** An electron is transferred from one atom to another.
   - **B** Two electrons are transferred from one atom to another.
   - **C** An electron is shared by two atoms.
   - **D** Two electrons are shared by two atoms.

8. A chemist is studying a molecule in which the atoms share electrons but do not share them equally. Which set of properties should the chemist expect the molecule to have?
   - **A** very low melting point, does not dissolve in water
   - **B** medium melting point, does not dissolve in water
   - **C** medium melting point, dissolves in water, solution does not conduct electricity
   - **D** very high melting point, dissolves in water, solution conducts electricity

9. When lithium is added to water, the two react, and the container holding the water becomes warmer. What does this indicate?
   - **A** The reaction is endothermic, and it absorbed more energy than it released.
   - **B** The reaction is endothermic, and it released more energy than it absorbed.
   - **C** The reaction is exothermic, and it absorbed more energy than it released.
   - **D** The reaction is exothermic, and it released more energy than it absorbed.

10. Which best describes what a catalyst does in a reaction?
    - **A** It increases reaction rate by lowering the activation energy.
    - **B** It increases reaction rate by adding heat to the reaction.
    - **C** It decreases reaction rate by keeping the reactants separate.
    - **D** It decreases reaction rate by absorbing energy from the reaction.

11. Which of the following is a decomposition reaction?
    - **A** $H_2O(l) \rightarrow H_2O(g)$
    - **B** $NaCl(aq) + AgNO_3(aq) \rightarrow NaNO_3(aq) + AgCl(s)$
    - **C** $O_2(g) + 2H_2(g) \rightarrow 2H_2O(l)$
    - **D** $2NH_3(g) \rightarrow N_2(g) + 3H_2(g)$

12. Which best describes a concentrated solution?
    - **A** a heterogeneous mixture with a lot of solute
    - **B** a homogeneous mixture with a lot of solute
    - **C** a heterogeneous mixture with little solute
    - **D** a homogeneous mixture with little solute

13. A diver jumps off an 8-meter board. Just as he jumps, he has a potential energy of 5,488 joules. What will his kinetic energy be when he is 2 meters above the water?
    - **A** 0 joules
    - **B** 1,372 joules
    - **C** 5,488 joules
    - **D** 9,604 joules

14. A book on the top bookshelf is 2 meters above the ground. It has 49 joules of potential energy. What is the mass of the book?
    - **A** 0.5 kg
    - **B** 2.5 kg
    - **C** 49.0 kg
    - **D** 98.0 kg

15. A hydroelectric power plant uses a waterfall with a 50-meter drop that flows at 2,500 kg/s. Which would be a better waterfall for the power plant to use?
    - **A** a waterfall with a 25-meter drop that flows at 4,500 kg/s
    - **B** a waterfall with a 30-meter drop that flows at 4,500 kg/s
    - **C** a waterfall with a 40-meter drop that flows at 2,000 kg/s
    - **D** a waterfall with a 45-meter drop that flows at 2,000 kg/s

16. Which is common among hydroelectric, wind, and fossil fuel energies?
    - **A** They are all renewable energy sources.
    - **B** They all convert potential energy to kinetic energy.
    - **C** They all have a step that depends on heat energy.
    - **D** They all involve turning a turbine, which turns a generator.

17. What is the difference between convection and conduction?
    - **A** Convection requires movement of matter; conduction does not.
    - **B** Convection involves the collision of atoms or molecules; conduction does not.
    - **C** Conduction requires movement of electrons; convection does not.
    - **D** Conduction involves the collision of electrons; convection does not.

18. What is the relationship between wavelength, frequency, and energy in the electromagnetic spectrum?
    - **A** As frequency increases, wavelength increases, but energy decreases.
    - **B** As frequency increases, wavelength and energy decrease.
    - **C** As wavelength increases, frequency increases, but energy decreases.
    - **D** As wavelength increases, frequency and energy decrease.

19. Which is an example of convection?
    - **A** The handle of a spoon stirring hot soup gets warm.
    - **B** Energy from the sun travels through space toward Earth.
    - **C** Warm air from an open oven rises to the ceiling and mixes with cool air.
    - **D** The floor in a room is warm where the sun hits it but is cool in the shadows.

20. Switches are used in electrical circuits to turn devices on and off. Which is the best description of how a switch works?
    - **A** It makes a break in the circuit to turn the device off.
    - **B** It sends the electric current in the opposite direction to turn the device off.
    - **C** It generates electricity to turn the device on.
    - **D** It stores electricity for the device to use when it is on.

21. A cyclist travels 150 meters in 20 seconds. What is the cyclist's average velocity?
    - **A** 7.5 m/s
    - **B** 130 m/s
    - **C** 170 m/s
    - **D** 300 m/s

22. A runner who was running at a velocity of 5.4 m/s accelerated to a velocity of 9.2 m/s in 2 seconds. What was the runner's acceleration?
    - **A** 1.7 m/s^2
    - **B** 1.9 m/s^2
    - **C** 3.4 m/s^2
    - **D** 3.8 m/s^2

23. A shopper pushes a cart with a net force of 41.5 N. The cart has a mass of 22.3 kg. What is the acceleration of the cart?
    - **A** 0.54 m/s^2
    - **B** 1.9 m/s^2
    - **C** 19.2 m/s^2
    - **D** 63.8 m/s^2

24. Which would have more momentum than a 6.3 kg bowling ball traveling at 4.1 m/s?
    - **A** a 7.2 kg bowling ball traveling at 3.1 m/s
    - **B** a 7.8 kg bowling ball traveling at 3.2 m/s
    - **C** a 5.7 kg bowling ball traveling at 4.6 m/s
    - **D** a 5.9 kg bowling ball traveling at 4.2 m/s

Use the following passage to answers questions 25–27.

Kendra is going to ride her bicycle to a store located up the hill from her house. As she starts to ride, she finds it difficult at first to turn the pedals to move the bicycle, but once she gets going, it is much easier. She comes to some steps and realizes that it would be too difficult to climb them on her bicycle, so she stops by squeezing the hand brakes, causing the brake pads to press against the sides of the tires. She looks around and sees a ramp, and she decides to head up the ramp to get to the store instead of taking the stairs.

25. What type of friction does Kendra need to overcome initially?
   - **A** fluid friction
   - **B** rolling friction
   - **C** sliding friction
   - **D** static friction

See page 115 for answers.

26. What type of force is used to stop the bicycle?
   - **A** static friction
   - **B** sliding friction
   - **C** magnetic force
   - **D** gravitational force

27. What simple machine does Kendra use instead of climbing the stairs?
   - **A** fulcrum
   - **B** inclined plane
   - **C** lever
   - **D** pulley

Lesson 1

1. B. An electrical wire needs to conduct, so it must be made from a metal. Gold is the only metal listed—it is to the left of the dividing line on the periodic table.

2. A. Chlorine is the only one in the same group as fluorine; elements in the same group have similar properties.

3. D. Sodium is on the left side of the periodic table, so it is a metal and has the properties of metals.

4. C. The element has some metallic properties and some non-metallic properties, so it must be a metalloid.

5. C. Elements in the same column are in the same group and have similar properties.

6. A. Argon is a non-metal, so it is non-conductive; it is inert, so it is not reactive; and shininess and malleability are properties of metals.

7. B. Atomic size increases as you go down a group, so neon would be smaller than argon; chlorine and hydrogen are not in the same group, and thus might not be inert.

Lesson 2

1. C. Covalent bonds form between non-metals; lithium, sodium, and aluminum are metals.

2. A. Cesium is a metal, and chlorine is a non-metal, so they transfer electrons, with the metal losing and the non-metal gaining.

3. C. Aluminum is a metal, so it will form a cation; oxygen is a non-metal so it will form an anion; the 2 indicates 2 Al's, and the 3 indicates 3 O's.

4. C. Linear, trigonal planar, and tetrahedral have no unbonded pairs of electrons.

5. D. Sharing the electrons means the compound is covalent; sharing them unequally means it is polar.

6. A. NaCl is ionic, so it probably has the highest melting point.

7. B. NaCl is ionic, and $C_6H_{12}O_6$ contains hydrogen and oxygen, so it is likely polar—both of those dissolve in water.

8. A. NaCl is ionic—only ionic compounds form water solutions that conduct electricity.

Lesson 3

1. C. A liquid, not a gas, is produced, and the liquid is not the same substance as the gases; atoms are not gained or lost in either physical or chemical reactions.

2. B. Combination reactions involve two or more reactants coming together to make a product; the other choices are decomposition reactions.

3. C. You need 6 carbon atoms on the reactant side because there are 6 carbon atoms on the product side.

4. The Law of Conservation of Matter.

5. Sample answer: Endothermic. The plants absorb energy from the sun, and that energy is stored in the glucose, so the products probably have more energy than the reactants.

6. Sample answer: They need an endothermic reaction because they want the reaction to absorb heat from the surrounding gel so that the gel cools.

7. Sample answer: Reaction B, because the energy of the products is higher than the energy of the reactants, and the additional energy is absorbed as heat from their surroundings.

8. Sample answer: They can add a catalyst to the gel or chemicals. It will lower the activation energy of the reactants, so the chemicals will need less energy to react.

Lesson 4

1. B. "Neutralize" means to make it neutral, or pH 7. The solution is acidic; adding water will raise the pH, but since water is neutral, the pH will still be below 7.

2. C. A and B are below −114, so it would be solid; D is above 79, so it would be gas.

3. A. The other three list ingredients, which are separated by the word and, while ammonium nitrate has a formula.

4. D. It is a mixture, so it can't be a compound. It separates into its parts, so it can't be a colloid or a solution.

5. B. 300 g is more sugar than can dissolve in 100 g water at 20°C, so stirring longer won't help. Adding more sugar or removing water will increase the sugar/water ratio, leaving her with even more undissolved sugar.

6. B. The substance started out as a gas, so the first flat part on the graph is where it condensed to a liquid; it is freezing from a liquid to a solid at the second flat part.

7. B. The substance is condensing, so it is changing from a gas to a liquid—both phases will be present.

8. Sample answer: The substance will be a liquid; −65°C is above the freezing point of −90°C and below the boiling point of 117°C.

Lesson 5

1. B. PE at the top will be the same as KE at the bottom. KE = $(1/2)mv^2$ = (0.5)(100 kg)$(16\ m/s)^2$ = 12800 kg m2/s^2. Since PE = mgh, h = PE/mg = (12800 kg m^2/s^2)/(100 kg)(9.8 m/s^2) = 13.06 m.

2. C. KE = mgh = (1 kg)(9.8 m/s^2)(100 m) = 980 kg m^2/s^2

3. D. They both convert into electricity the kinetic energy of the turbines they turn; electricity is another form of kinetic energy.

4. A. The potential energy in the gasoline is released by the reaction of burning, and the resulting movement of the engine is kinetic energy.

5. B. The kinetic energy of the light changes to electrical energy when the current flows to the cell phone battery, where it is stored as potential energy.

6. C. The water that is pumped uphill gains potential energy, so they are using excess kinetic energy when they don't need it to make sure they have extra potential energy when they do need it.

Lesson 6

1. C. Radiation transfers energy without contact, conduction transfers energy by contact from high to low areas, and convection transfers energy by the flow of hot fluids.

2. B. Convection requires contact between materials, so the higher energy molecules can collide with lower energy molecules to transfer the energy.

3. A. Higher frequency (lower wavelength) radiation has more energy.

4. D. UV radiation has a higher frequency than infrared and visible. The fact that you can see through sunscreen and get heat from the sun indicates that only the UV is absorbed by the sunscreen.

5. electric, heat

6. incomplete

7. electrons

8. stores

Lesson 7

1. C.

2. B. $v = d/t$ = 100 m/9.63 s = 10.4 m/s

3. A. a = $(v_{end} - v_{begin}$ $/t$ = (8.5 m/s − 0 m/s)/2 s = 4.25 m/s^2

4. C. $F = ma$ = (65 kg)(5.5 m/s^2) = 357.5 N

5. B. 115 N − 25 N − 27 N = 63 N

6. D. $p = vm$ = (9.8 m/s)(65 kg) = 637 kg m/s

Lesson 8

1. B. Rolling friction is shown in the image. The person is pushing a ball, which has a round surface.

2. D. Static friction is present because the box is not moving.

3. D. In part 4, the person is exerting a force that is causing the box to move.

4. D. The person is pushing the box up a ramp, which is an example of an inclined plane.

5. A. Work is dependent on the amount of force required to move an object a distance, so if the amount of work stays the same, the greater the distance traveled, the less force required.

6. C. The magnet falls to the floor because of the gravitational force Earth exerts on the magnet.

7. B. A lever pivots about a fixed point called a fulcrum.

8. B. A moveable pulley allows the distance that the object is moved to increase, thereby decreasing the amount of force required.

Unit Practice Test

1. D.
2. A.
3. A.
4. C.
5. B.
6. B.
7. D.
8. C.
9. D.
10. A.
11. D.
12. B.
13. B.
14. B.
15. B.
16. D.
17. A.
18. D.
19. C.
20. A.
21. A.
22. B.
23. B.
24. C.
25. D.
26. B.
27. B.

- **acceleration** – rate of change of velocity of an object
- **acid** – a substance that lowers the pH of water when added to it
- **activation energy** – energy necessary for two colliding atoms or molecules to react
- **anion** – an atom with a negative charge as a result of having gained electrons
- **atom** – the smallest particle of matter that still has the properties of the element that makes up the matter
- **base** – a substance that raises the pH of water when added to it
- **batteries** – energy stored in chemical reactions that is released as electrical energy
- **boiling point** – temperature at which a substance changes from the liquid state to the gas state
- **catalyst** – a substance that speeds up a reaction by lowering activation energy; is not used up in the reaction
- **cation** – an atom with a positive charge as a result of having lost electrons
- **chemical bond** – attractive force between atoms in a compound, formed by transferring or sharing electrons
- **chemical change** – a change that results in one or more new substances being produced
- **chemical equation** – shows the formulas, states of matter, and proportions of products and reactants in a reaction
- **chemical formula** – representation of the ratio of elements in a compound, using the element symbols and numerical subscripts
- **chemical reaction** – when the bonds of one or more reactants are broken, atoms rearranged, and bonds formed, resulting in new products
- **circuit** – a closed loop containing at least one power source and one device connected by wires
- **colloid** – heterogeneous mixture that does not settle or separate on its own
- **combination reaction** – two or more elements or compounds reacting to form one compound (also known as synthesis reaction)
- **compound** – a pure substance made by chemically bonding atoms in fixed ratios
- **condense** – to change from the gas state to the liquid state
- **conduction** – energy movement from one location to another by contact
- **convection** – energy movement from one location to another by movement of liquid or gas
- **covalent bond** – attractive force between atoms sharing two electrons
- **current** – flow of electric charge; can be carried by electrons moving through a circuit
- **decomposition reaction** – one compound reacting to form two or more elements or compounds
- **distance** – measurement of the space between two points
- **double bond** – a covalent bond in which four electrons are shared
- **electricity** – energy movement associated with an electrical current
- **electromagnetic spectrum** – the range of all frequencies' electromagnetic radiation, such as microwaves and visible light
- **electron** – a small, negatively charged particle in an atom, found outside the nucleus
- **electron cloud** – the region outside the nucleus in which electrons of different energy are likely to be found
- **electrostatic attraction** – attraction between positive and negative charges of cations and anions
- **element** – a pure substance composed entirely of the same type of atom, represented by a symbol on the periodic table
- **endothermic reaction** – a reaction in which more heat is absorbed than is released; lowers the temperature of its surroundings
- **exothermic reaction** – a reaction in which more heat is released than is absorbed; raises the temperature of its surroundings
- **fluid friction** – the resistance of an object's movement through a liquid or gas
- **force** – a push or a pull; equal to the mass of an object times the acceleration acting on it

- **freeze** – to change from the liquid state to the solid state
- **frequency** – the number of cycles of a wave of electromagnetic radiation that pass a point in one second
- **friction** – the resistance or opposition to motion
- **fulcrum** – the pivot point of a lever
- **gas** – state of matter that takes the shape and volume of its container
- **gravitational force** – the force of attraction between any two objects with mass
- **heterogeneous** – matter with non-uniform composition
- **homogeneous** – matter with uniform composition
- **hydroelectric** – electric power generated by turbines moved by water flowing downward
- **inclined plane** – a simple machine consisting of a ramp
- **ionic bond** – attractive force between atoms as a result of electrons transferring from one to the other
- **kinetic energy** – energy of motion
- **Law of Conservation of Energy** – energy cannot be created or destroyed
- **Law of Conservation of Matter** – matter (or mass) cannot be created or destroyed
- **lever** – a simple machine that consists of a bar that pivots about a point
- **light energy** – energy contained in electromagnetic waves
- **liquid** – state of matter that takes the shape but not the volume of its container
- **magnetic force** – the force exerted between magnets
- **mass** – the amount of matter in an object
- **melting point** – temperature at which a substance changes from the solid state to the liquid state
- **metalloids** – having properties of both metals and non-metals; the elements between metals and non-metals on the periodic table: B, Si, Ge, As, Sb, Te, At
- **metals** – elements to the left of the metalloids on the periodic table; conductive, malleable, and shiny
- **mixture** – matter containing more than one pure substance
- **molecule** – smallest particle of a compound that still has the properties of that compound; has atoms bonded together in the ratio given by the compound's formula
- **momentum** – a quantity of motion; equal to the mass of an object times its velocity
- **net force** – the sum of forces acting on an object; forces in the same direction are added, in opposite directions are subtracted
- **neutron** – a neutral particle in an atom, found inside the nucleus
- **non-metals** – elements to the right of the metalloids on the periodic table; non-conductive, brittle, and dull
- **non-polar** – a molecule in which the electrons are shared equally through covalent bonds
- **nuclear force** – the attractive force between protons and neutrons in the nucleus
- **nuclear power** – electric power generated by turbines moved by steam generated by the heat of nuclear reactions
- **nucleus** – central part of the atom; contains most of the atom's mass, as protons and neutrons
- **Periodic Law** – when elements are arranged in order by number of protons, groups with similar properties occur periodically (such as every 8 elements)
- **periodic table of elements** – arrangement of the elements in order by number of protons in rows and columns, with each column of elements having similar properties
- **pH scale** – scale of acidity from 0 to 14, with 0 being most acidic and 14 being least acidic (most basic)
- **physical change** – a change that does not change the identity of the substance
- **polar** – a molecule in which the electrons are shared unequally through covalent bonds, producing slightly positive and negative poles
- **potential energy** – stored energy; energy that can be released by changing it into another form
- **product** – a compound or element that is the result of a chemical reaction

Unit Glossary

- **proton** – a positive particle in an atom, found inside the nucleus
- **pulley** – a simple machine consisting of a grooved wheel through which a rope passes
- **radiation** – energy movement through space in the form of waves or particles
- **reactant** – a compound or element that is an ingredient in a chemical reaction
- **reaction rate** – how quickly or slowly a reaction progresses
- **rolling friction** – the resistance of a round object's movement
- **sliding friction** – the resistance to motion that exists between two objects sliding past one another
- **solar** – electric energy generated by light energy from the sun moving electrons through solar panels
- **solid** – state of matter that takes neither the shape nor the volume of its container
- **solute** – the component in a solution that is dissolved
- **solution** – a homogeneous mixture of two or more pure substances
- **solvent** – the component in a solution that does the dissolving
- **static friction** – the resistance of motion that exists between a stationary object and the surface on which it sits
- **suspension** – a heterogeneous mixture that settles or separates on its own
- **synthesis reaction** – two or more elements or compounds reacting to form one compound (also known as combination reaction)
- **unshared electrons** – pairs of electrons in the outermost level of an atom that are not shared in a covalent bond with another atom
- **velocity** – the speed and direction of a moving object
- **wavelength** – length from peak to peak of a wave of electromagnetic radiation
- **wind power** – electric power generated by turbines moved by the wind

Study More!

Consider exploring these concepts, which were not introduced in the unit:

- Atomic mass and mass number
- Isotopes and radioactivity
- Allotropes
- Nuclear changes
- Substitution reactions
- Aeration of liquids
- Filtering of suspensions and colloids
- Polymers
- Plastics
- Distillation
- Plasma (a state of matter)
- Common settings for different energy units
- Entropy
- Geothermal energy
- Alternating current
- Direct current
- Wave interference
- Doppler effect
- Magnetic fields
- Action
- Displacement
- Pressure
- Resistance
- Power

Life Science

Life science is the study of living things, including plants, animals, and bacteria. It examines how living things work and how they interact with each other and with the environment.

Understanding the living world is important for many aspects of modern life. Knowledge of how your own body functions can help you make better decisions about your health. The living world also provides food, clean water, and resources that people need to survive, and studying life science will help you understand important issues facing the world.

KEY WORDS

- abiotic component
- active transport
- adaptation
- allele
- amino acid
- ATP
- behavior
- biome
- biotic component
- carbon dioxide
- cell membrane
- cell wall
- cellular respiration
- chlorophyll
- chloroplast
- chromosome
- circulatory system
- cladogram
- cold-blooded
- common ancestor
- cytoplasm
- decomposers
- diffusion
- digestive system
- disease
- diversity
- DNA
- domain
- ecosystem
- endocrine system
- endoplasmic reticulum
- environmental disease
- enzymes
- esophagus
- eukaryote
- evolution
- excretory system
- fertilization
- food chain
- food web
- fossil record
- gametes
- gene
- genetics
- genotype
- germinate
- Golgi apparatus
- heredity

UNIT 4

Life Science

KEY WORDS

- immunity
- immunization
- immunodeficiency
- inflammation
- instinct
- kingdom
- leaves
- meiosis
- messenger RNA
- metabolism
- minerals
- mitochondrion
- mitosis
- muscular system
- mutation
- natural selection
- nervous system
- nonvascular plant
- nucleus
- nutrients
- nutrition
- organelle
- osmosis
- oxygen
- passive transport
- pathogen
- phenotype
- phloem vessel
- photosynthesis
- pistil
- pollen
- pollination
- prokaryote
- protein
- reflex
- replication
- respiratory system
- ribosome
- RNA
- roots
- seed
- sense organs
- skeletal system
- species
- stamen
- stem
- symbiosis
- symmetry
- taxonomy
- traits
- transcription
- transfer RNA
- translation
- trophic levels
- urine
- vascular plant
- voluntary muscles
- warm-blooded
- white blood cells
- xylem

Although most cells are too small to be seen without a microscope, they are the basic units that make up living things. All living organisms contain one or more cells, and in organisms with more than one cell (multicellular), cells can be specialized into different types.

KEY WORDS

- cell membrane
- cell wall
- chloroplast
- chromosomes
- cytoplasm
- endoplasmic reticulum
- eukaryote
- Golgi apparatus
- mitochondrion
- nucleus
- organelles
- prokaryote

Types of Cells

A **eukaryote** is an organism whose cells contain a nucleus and other **organelles** that are contained within membranes. The nucleus of a eukaryotic cell contains the cell's genetic material, or DNA, organized into bundles called **chromosomes**. Eukaryotes may have one or many cells. Plants, animals, and fungi are examples of eukaryotes.

A **prokaryote** is a single-celled organism that does not contain a nucleus or other organelles, and its DNA is found in its cytoplasm. Bacteria are a type of prokaryote.

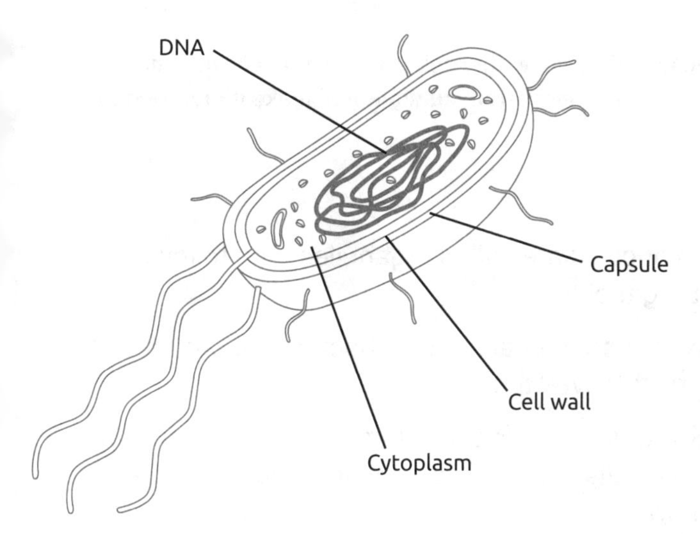

Bacteria and other prokaryotic cells have no internal membranes or organelles. Their DNA is not contained in a nucleus.

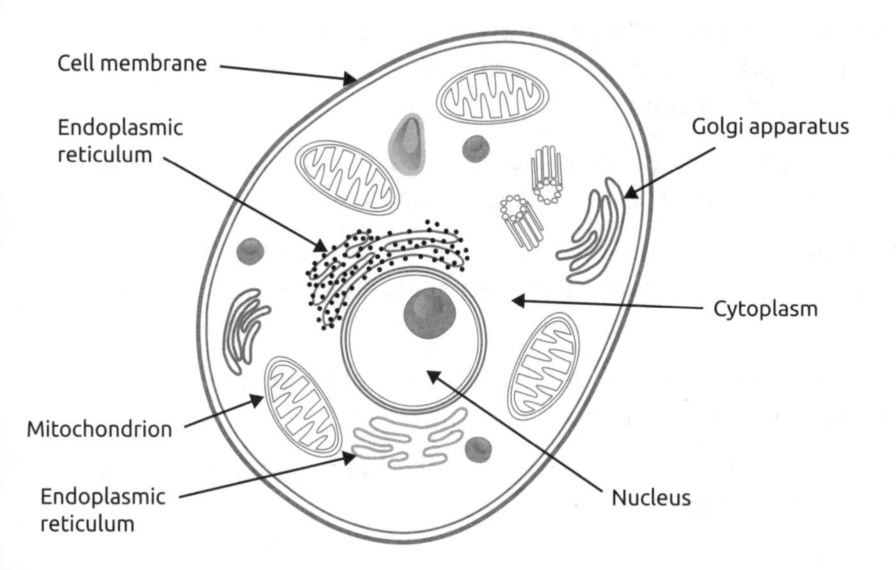

Animal cells contain organelles with membranes and a nucleus, usually found in the center of the cell. Because they do not have cell walls, animal cells often have irregular shapes and can change shape.

Cells

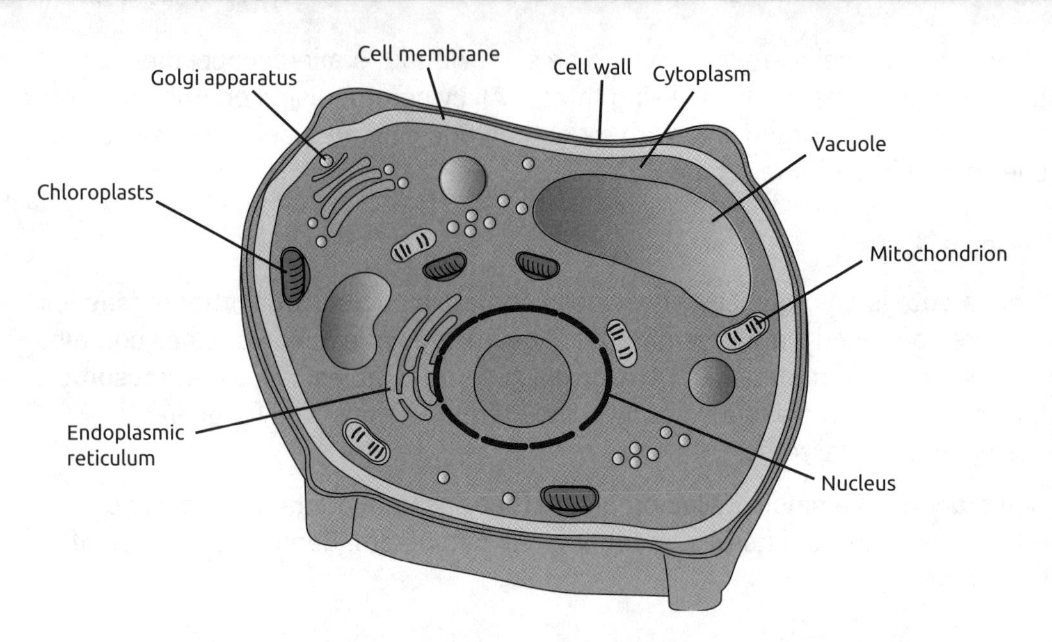

Plant cells have fixed shapes because they have rigid cell walls. Plant cells contain chloroplasts, organelles that capture the energy from sunlight and make it into food for the cell.

Parts of a Cell

The working parts of a eukaryotic cell are called organelles. The following are some of the most important organelles:

Cell membrane: the outer layer of an animal, fungus, bacteria, or protist cell, which determines what enters or leaves the cell.

Cell wall: a tough outer layer that surrounds the cell membrane in some types of cells. Plant cells have rigid cell walls that help to provide structure to the plant. Animal cells do not have cell walls.

Chloroplast: an organelle found in plant cells that captures the energy from sunlight and uses it to make food for the cell.

Cytoplasm: the contents of a cell within the cell membrane and outside of the nucleus. The cytoplasm contains the organelles, surrounded by a gel-like material made up mostly of water.

Endoplasmic reticulum: a system of interconnected, folded sacs and tubes that produces materials, especially proteins, and transports them through the cell.

Golgi apparatus: an organelle that packages and sorts molecules for transport through the cell. In most eukaryotes, the Golgi apparatus resembles a stack of flat discs.

Mitochondrion: the cell's "powerhouse," a structure that provides energy to the cell by converting food into ATP, the cell's energy source.

Nucleus: the control center of the cell, which contains the genetic material (DNA) that directs the activities of the cell.

Complete the activities below to check your understanding of the lesson content.

Skills Practice

Identify each organelle based on the description of its function.

1. The ____________________ separates the contents (cytoplasm) of an animal cell from the outside of the cell.

2. The ____________________ converts food into energy for the cell.

3. The ____________________ captures energy from the sun and produces food for the cell.

4. The ____________________ packages and sorts molecules, such as proteins, for transport through the cell.

Answer the following questions based on the content covered in the lesson.

5. Name two features found in eukaryotic cells that are not found in prokaryotic cells.

____

____

6. A crime scene investigator finds some cells on the shoes of a crime suspect and views the cells under a microscope. She observes that all of the cells are similar in shape and size, have rigid cell walls, and appear to contain chloroplasts. The investigator wants to ask an expert to identify the cells, in order to learn more about where the suspect has been. What type of scientist should she ask?

   A a biologist who studies animals
   B a biologist who studies plants
   C a biologist who studies insects
   D a biologist who studies bacteria

See page 171 for answers and help.

KEY POINT!

This lesson focuses on the differences between various types of cells. However, all cells share some basic structures and functions. For example, all cells have DNA, and all cells use energy. Eukaryotic cells, such as plant and animal cells, share even more features, such as a nucleus and mitochondria.

TEST STRATEGY

When a question includes a long description, it can be helpful to underline the key words and phrases in the question. Try to find and underline the scientific observations described in the question. Also look for words or phrases that relate to one of the answer choices.

Cell Functions

KEY WORDS

- active transport
- amino acids
- chromosome
- diffusion
- DNA
- gametes
- meiosis
- messenger RNA
- mitosis
- osmosis
- passive transport
- protein
- replication
- ribosome
- RNA
- transcription
- transfer RNA
- translation

All living things must carry out certain activities in order to survive. Like single-celled organisms, the cells of our bodies are able to accomplish all of the functions necessary for life. For instance, cells must obtain nutrients and oxygen, remove waste, and reproduce.

Transport in Cells

Substances move into and out of the cell in two main ways: active and passive transport.

Passive transport involves the movement of materials into and out of cells without the use of energy. **Diffusion** is a type of passive transport in which materials move from an area of high concentration to one of low concentration. This is similar to the process by which an odor can fill a space. **Osmosis** is the diffusion of water across the cell membrane.

In **active transport**, energy is required to move particles across the cell membrane. Active transport can be accomplished in a number of ways, including the use of special proteins in the cell membrane.

Reproduction in Cells

Living things must reproduce, and they must ensure that the new cells are able to do everything that the parent cells can do. **DNA** (deoxyribonucleic acid) contains the code for the ways in which life activities are accomplished. DNA contains the instructions for the production of proteins, which help carry out most of these functions, from breaking down nutrients to reproduction. The **chromosomes** of the cell, found in the nucleus, contain the individual's DNA. The process by which the DNA of chromosomes is copied is called **replication**.

Cells reproduce by making copies of themselves; they do not require other cells, and the new cells are identical to the parent cells. Eukaryotic cells generally reproduce through **mitosis**, a series of steps in which the chromosomes of a cell are copied and the nucleus and cell are split, resulting in two new cells that are exact replicas of the parent cell. The new cells have the same DNA and the same number of chromosomes.

Meiosis: A Special Type of Cell Reproduction

The first step in sexual reproduction is **meiosis**. Meiosis is a process similar to mitosis, but it includes two separate divisions of the cell. During the initial step of meiosis, the chromosomes are copied and then split so that rather than each cell having a pair of chromosomes, each cell has a single chromosome. In the second division, the chromosomes are divided again, and the result is four cells, each with half the number of chromosomes as the original cell. These newly formed cells are called **gametes**; the male gamete is the sperm, and the female gamete is the egg.

Protein Synthesis

Although DNA contains the genetic code, a different but similar molecule, **RNA**, carries that information from the nucleus to the cytoplasm, where proteins are produced. RNA is formed using the DNA sequence of bases as a template through a process called **transcription**. The RNA then leaves the nucleus and travels to an organelle called the **ribosome**, which is the site of protein synthesis. This type of RNA is called **messenger RNA** (mRNA) because it delivers the DNA's "message" from the nucleus to the cytoplasm.

In the cytoplasm, a different type of RNA carries amino acids to the ribosome. It is called **transfer RNA** (tRNA) because it transfers amino acids. **Amino acids** are the building blocks of proteins. As the tRNA matches up to the appropriate mRNA region (one with a complementary code), it leaves behind amino acids, which then bond together. The order in which the amino acids bond is determined by the sequences of bases in the messenger RNA. The amino acids continue to bond together to form a chain that will become a **protein**. This process, whereby the genetic code is translated to produce the appropriate sequence of amino acids for a specific protein, is called **translation**.

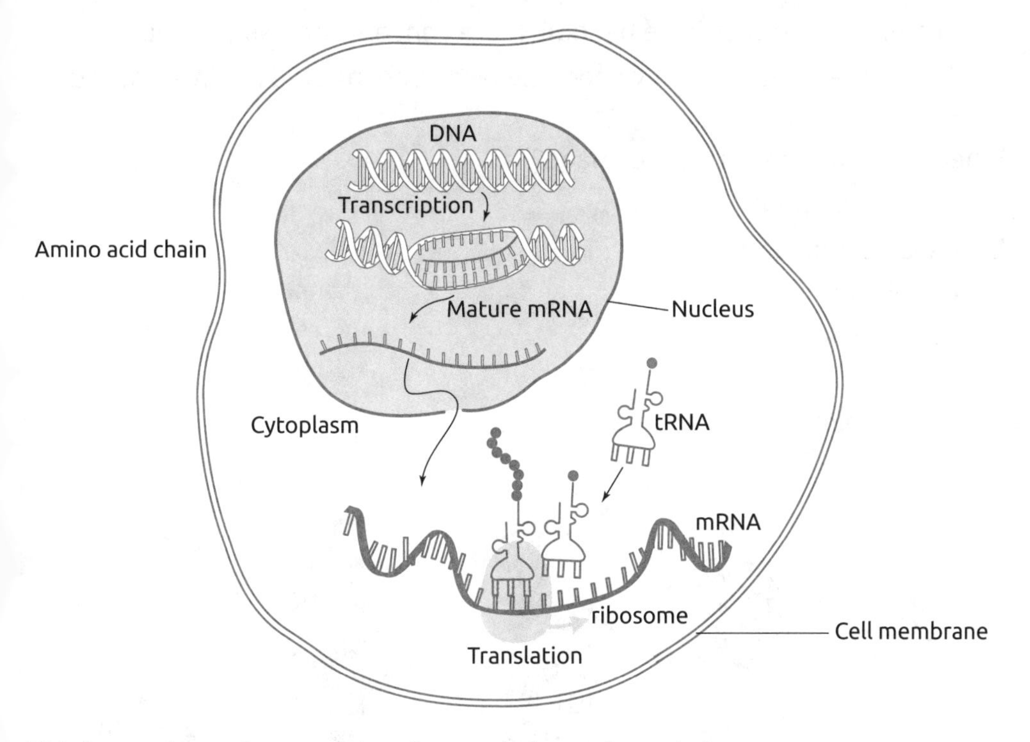

This image shows the processes of transcription and translation.

Lesson Practice

Complete the activities below to check your understanding of the lesson content.

Skills Practice

Use the image below to answer questions 1 and 2.

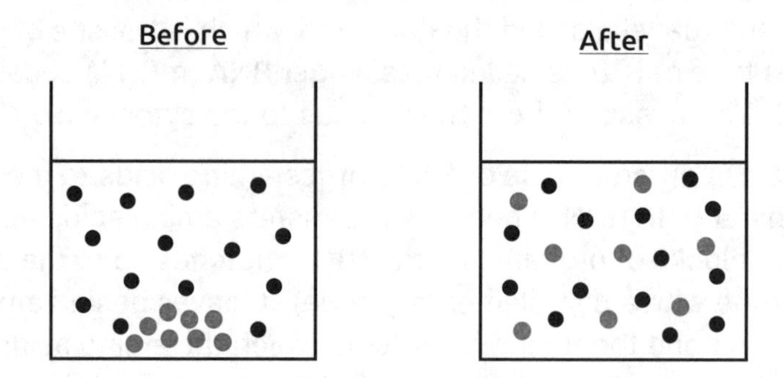

1. Which best describes the process occurring in the image?
   - **A** The gray molecules are diffusing through the substance.
   - **B** The purple molecules are diffusing through the substance.
   - **C** The gray molecules have become more concentrated as time passed.
   - **D** The purple molecules have become less concentrated as time passed.

2. What process is shown in the image?
   - **A** active transport
   - **B** passive transport
   - **C** translation
   - **D** transcription

Use the image below to answer questions 3–6.

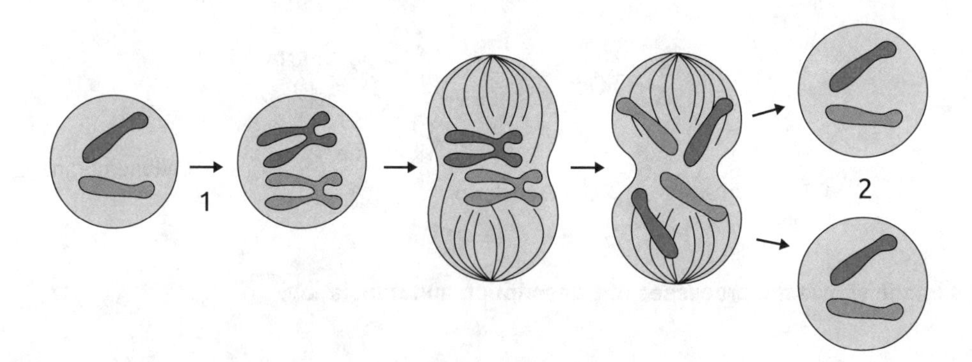

3. What is happening at point 1 in the image?
   - **A** diffusion
   - **B** replication
   - **C** transcription
   - **D** translation

KEY POINT!

Remember that energy is required for passive transport but not for active transport.

TEST STRATEGY

Look for any familiar terms and think about or jot down their meanings. For any unfamiliar terms, try to break down the words, using prefixes, suffixes, stems, and cognates to determine meaning.

4. Which term best describes the process shown in the image?
   - **A** meiosis
   - **B** mitosis
   - **C** active transport
   - **D** protein synthesis

5. Which is true regarding the two cells at point 2 in the image?
   - **A** The two cells will fuse together.
   - **B** The two cells will split apart again.
   - **C** Each cell is identical to the parent cell.
   - **D** Each cell has half as many chromosomes as the parent cell.

6. Which of the following are NOT produced in the process depicted in the image?
   - **A** skin cells
   - **B** heart cells
   - **C** muscle cells
   - **D** reproductive cells

7. What term best describes the formation of RNA using the code from a strand of DNA?
   - **A** diffusion
   - **B** replication
   - **C** transcription
   - **D** translation

8. Which is the process whereby amino acids are brought to the ribosome and bound together in a specific sequence determined by the genetic code?
   - **A** diffusion
   - **B** replication
   - **C** transcription
   - **D** translation

See page 171 for answers and help.

Plants

KEY WORDS

- fertilization
- germinate
- leaves
- nonvascular plants
- phloem vessels
- photosynthesis
- pistil
- pollen
- pollination
- roots
- seed
- stamen
- stem
- vascular plants
- xylem

It is difficult to imagine looking around and not seeing plants; you see them growing up from the Earth in many different forms, from tall pine trees to beautiful flowers and soft grass. Like humans, plants must accomplish all of life's activities, including growth, circulation of materials, and reproduction.

Plant Classification

Plants can be allocated into two principal groups: vascular and nonvascular plants. **Vascular plants**, such as ferns, flowering plants, and trees, contain vascular tissue, or tubes, for transport of materials; they also have true roots, leaves, and stems.

Nonvascular plants, such as mosses and liverworts, do not contain vascular tissue and lack true roots, stems, and leaves. Since they do not have any vessels for transport, they grow low to the ground, usually in moist areas.

Plant Structures

If you were to remove a plant carefully from the soil, you would be able to observe the main structures of a vascular plant—the leaves, roots, and stem. The functions of the **roots** are to anchor the plant in the ground and to absorb water and minerals that travel through the plant via the **xylem,** vascular tissue that transports water and minerals to the leaves. The **leaves** are the sites of **photosynthesis,** the process that allows a plant to make its own nutrients using energy from the sun and carbon dioxide. Nutrients are conveyed to other parts of the plant for use or storage via the **phloem vessels**. The xylem and phloem run through the **stem** to the roots and leaves of the plant.

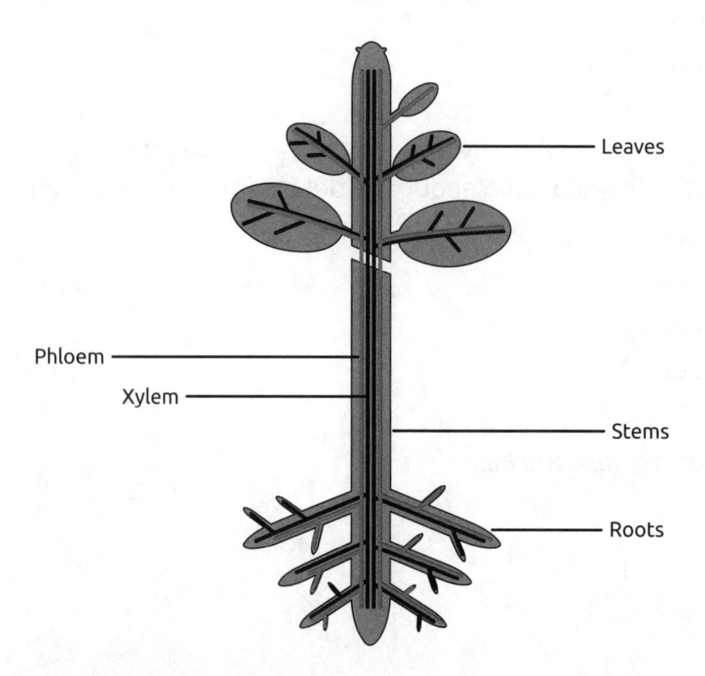

Plant structures

Plant Reproduction

Plants can reproduce asexually or sexually. Sexual reproduction occurs in most plants, including trees and flowering plants. Sexual reproduction involves the fusion of male and female gametes and results in the production of a new individual plant, similar to both parents but not identical to either. Meiosis, the process by which the gametes are produced, occurs in the flowers. Some flowers contain both male and female reproductive organs, while others contain only one or the other. The male reproductive organ is the **stamen**, which is made up of the anther and filament.

Pollen grains contain the male gametes and are found on the tip of the anther. The female reproductive structure is the **pistil**. The pistil includes the ovary, where the ovules, the female gametes, are found.

The method by which pollen is transferred from the anther to the pistil is called **pollination**, which can occur through the actions of birds, bees, and the wind. After a flower is pollinated, the pollen grains travel down to the ovule, where **fertilization**, the process by which male and female gametes fuse, happens. As a result of this union, a **seed** develops.

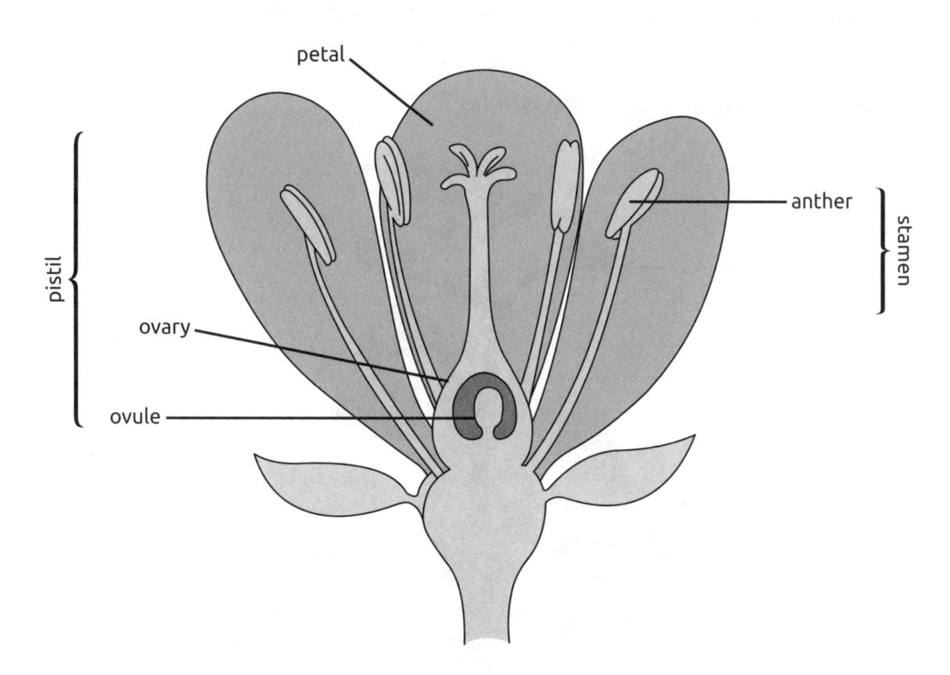

The structures involved in sexual reproduction in plants

Plant Growth

Under appropriate temperature and moisture conditions, a seed will **germinate**, or begin to grow. The plant uses nutrients and energy stored in the seed to grow, to develop into a plant with roots, leaves, and a stem, and to begin producing food on its own.

Plant growth generally occurs in specific areas. Development in the tips of the roots and stems accounts for an increase in the height of a plant, while growth areas in the spaces between the xylem and phloem result in an enlargement of a plant's width.

Lesson Practice

KEY POINT!

Remember that vascular plants have vessels that transport materials to where they are needed.

KEY POINT!

Remember that water and minerals are taken into a plant from the soil, not absorbed through leaves.

TEST STRATEGY

Put the question in your own words to make sure you understand it.

Complete the activities below to check your understanding of the lesson content.

Skills Practice

Answer the questions based on the content covered in the lesson.

1. Which of the following uses photosynthesis to obtain nutrients but has no phloem through which to transport the nutrients?
   A moss
   B grass
   C apple tree
   D rose bush

2. Which explains why xylem can be found in a plant's roots?
   A Xylem produces minerals in a plant's roots.
   B Xylem is involved with the growth of a plant's roots.
   C Xylem moves sugar from the leaves to the roots of a plant.
   D Xylem transports water from the roots to the leaves of a plant.

Use the diagram below to answer questions 3–5.

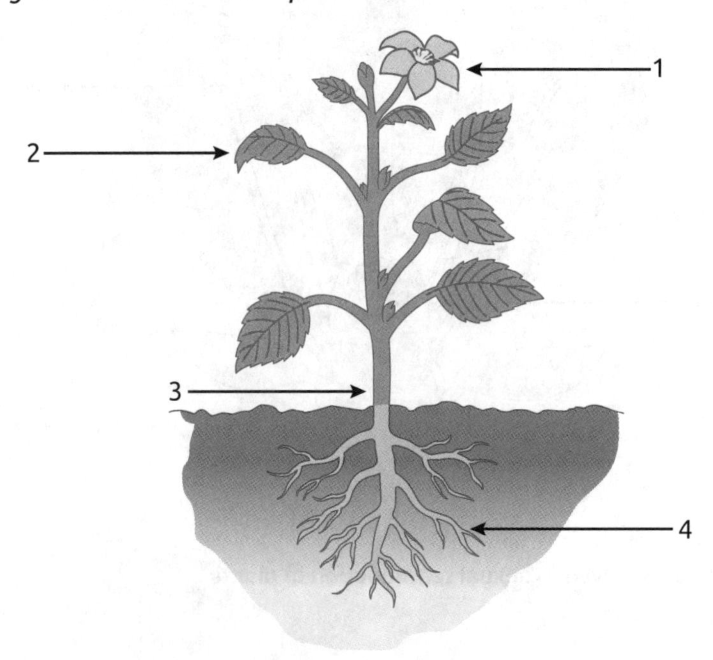

3. Where does plant reproduction take place?
   A 1
   B 2
   C 3
   D 4

4. Which part of the plant makes food for its survival?
   - **A** 1
   - **B** 2
   - **C** 3
   - **D** 4

5. Which part of the plant takes in water and minerals?
   - **A** 1
   - **B** 2
   - **C** 3
   - **D** 4

Use the description below to answer questions 6 and 7.

One summer day, a bee is seen moving from flower to flower on a lily plant. Several days later, the plant drops seeds to the ground. The next spring, new lily plants are growing where the bee had once been seen.

6. How did the bee help in the reproduction of the lily plant?
   - **A** The bee moved pollen from the anther to the pistil of the plant.
   - **B** The bee moved pollen from the pistil to the anther of the plant.
   - **C** The bee fertilized the anther as it moved from flower to flower.
   - **D** The bee fertilized the pollen as it moved from flower to flower.

7. After the seed dropped to the ground, when conditions were just right, the seed _______.
   - **A** germinated
   - **B** fertilized
   - **C** pollinated
   - **D** vascularized

See page 171 for answers and help.

Plant Metabolism

KEY WORDS

- ATP
- carbon dioxide
- cellular respiration
- chlorophyll
- chloroplasts
- metabolism
- oxygen
- photosynthesis

Every year, roughly 46,000 to 58,000 square miles of forests are lost due to deforestation, which equates to about 36 football fields every minute. Plants play an important role in our environment, acting as a carbon sink; they take in carbon dioxide from the air that would otherwise stay in the atmosphere and contribute to global warming. What do the plants do with this carbon dioxide?

A Plant's Leaves: Where Photosynthesis Occurs

A plant's leaves contain structures called stomata, tiny holes that allow **carbon dioxide** (CO_2) in and **oxygen** (O_2) out. Also within the leaf are veins that transport water and mesophyll cells. Within the mesophyll cells are organelles called **chloroplasts**, which contain **chlorophyll**, the substance that gives plants their green color and absorbs light energy from the sun.

Photosynthesis: The Light Reaction

Plants perform many chemical reactions within their cells that allow them to carry out life functions. Together, these interrelated reactions are referred to as **metabolism**. One of the chemical reactions involved in a plant's metabolism is **photosynthesis**. Photosynthesis is the process in which plants (and some algae) convert light energy to chemical energy, which is stored in the form of sugar. It is represented by the following chemical formula:

$6CO_2 + 6H_2O$ (+ light energy) $\rightarrow C_2H_{12}O_6 + 6O_2$

Photosynthesis has two parts: the light reaction and the dark reaction. In the light reaction, sunlight enters the chloroplast and is absorbed by chlorophyll, commencing a chain of reactions. These reactions split the water (H_2O) molecules and result in the production of ATP, NADPH, and oxygen. **ATP** is a molecule that all cells can easily use as energy for life functions, and NADPH is a molecule that acts as an electron carrier. Oxygen is released into the atmosphere through the stomata, and the ATP and NADPH are used in the dark reaction.

Photosynthesis: The Dark Reaction

The dark reaction (also known as the Calvin Cycle) is called "dark" because it does not use light energy. In this series of reactions, first, carbon dioxide is combined with a sugar called RuBP. Then, using the ATP and NADPH from the light reaction, the carbon dioxide and RuBP combination is broken apart, and the pieces are converted into a sugar called glucose.

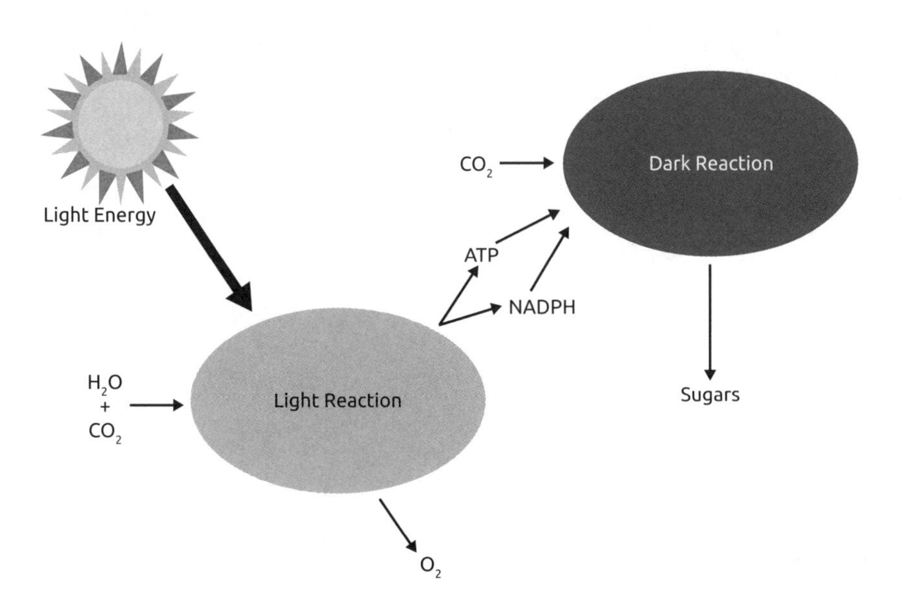

Photosynthesis consists of the light and dark reactions. Carbon dioxide, water, and the sun's energy are the reactants, and oxygen and glucose are the products.

Cellular Respiration

Cellular respiration occurs in both plants and animals within the mitochondria of cells. It transfers the energy stored in glucose into molecules of ATP. Cellular respiration can be represented by the following formula:

$C_6H_{12}O_6 + 6O_2 \rightarrow 6CO_2 + 6H_2O + ATP$

Cellular respiration occurs in two steps: anaerobic respiration and aerobic respiration.

Anaerobic Respiration (Glycolysis)

During this part of cellular respiration, the plant breaks down glucose into pyruvate and also releases ATP and NADH. This occurs in the cytoplasm and does not require the presence of oxygen.

Aerobic Respiration

Aerobic respiration occurs in the mitochondria and requires oxygen. During this stage of respiration, pyruvate is broken down and rearranged several times in a cycle (the Krebs Cycle). This produces carbon dioxide and energy in the form of NADH, FADH, and ATP. Electrons are pulled from this step and passed into an electron transport chain, which releases a large amount of energy used to make ATP. At the end of the electron transport chain, water is formed when the electrons join with oxygen.

Complete the activities below to check your understanding of the lesson content.

Skills Practice

Answer the questions based on the content covered in the lesson.

1. Which molecules are produced from photosynthesis?
   - **A** O_2 and CO_2
   - **B** H_2O and CO_2
   - **C** H_2O and $C_6H_{12}O_6$
   - **D** O_2 and $C_6H_{12}O_6$

2. What is the function of oxygen in cellular respiration?
   - **A** to provide the energy needed to break down pyruvate
   - **B** to combine with light energy and water to produce glucose
   - **C** to pull electrons along the electron transport chain and form water
   - **D** to aid in the breakdown of carbon dioxide and begin photosynthesis

3. In hot, dry weather, some grasses will close their stomata in order to reduce the amount of water lost to the atmosphere. What would be a result of this action?
   - **A** The grass will carry out only aerobic respiration.
   - **B** The grass will cease photosynthesis and turn brown.
   - **C** The grass will pull in carbon dioxide from its roots.
   - **D** The grass will increase its output of oxygen and glucose.

4. Why is photosynthesis important to animals?

   ____

   ____

   ____

5. Mitochondria are often referred to as the "powerhouses" of cells. How does cellular respiration explain why they are given this name?

   ____

   ____

   ____

See page 171 for answers and help.

KEY POINT!

Cellular respiration can be thought of as the opposite of photosynthesis. Photosynthesis uses carbon dioxide and water to produce oxygen and glucose. Cellular respiration uses oxygen and glucose to produce carbon dioxide and water.

TEST STRATEGY

Cover the answers to the question. Think of what you would write if the answers weren't there, and then uncover the answers. Choose the one that best matches your own answer.

Animals

KEY WORDS

- behavior
- cold blooded
- instinct
- reflex
- symmetry
- warm blooded

In 2011, scientists estimated that there were approximately 7.77 million species of animals, of which 953,434 were classified. The process of classifying organisms can be difficult work, relying on a number of factors, including development, genetics, and fossils. Placing animals into the main groups, however, is pretty straightforward.

Characteristic Traits

Animals can be separated into larger groups based on certain physical characteristics, such as **symmetry**. Animals with radial symmetry have identical parts arranged in a circle around a central axis. Animals with bilateral symmetry are the same on two sides. Another characteristic trait is how an organism maintains its body temperature. Animals that are **cold blooded** rely on the environment for their internal temperature. Reptiles can often be seen sunning themselves, because this raises their body temperature. Mammals are **warm blooded**, which means they maintain an internal temperature. For humans, our internal temperature is 98.6 degrees Fahrenheit regardless of the temperature outside.

Invertebrates and Vertebrates

	Group	Characteristics	Examples
Invertebrates: no backbone, no spinal cord	Porifera	Simple animals with no distinct tissues or organs and no symmetry	Sponge
	Cnidarians	Radial symmetry, stinging cells, tissues, no organs	Jellyfish
	Mollusks	Bilateral symmetry, organs, soft body, distinct head and foot	Clam, squid
	Annelids	Segmented body, bilateral symmetry, closed circulatory system	Earthworm
	Arthropods	Segmented body, bilateral symmetry, exoskeleton (outer skeleton)	Spider, crab, grasshopper
Vertebrates: internal skeleton, backbone, spinal cord	Fish	Scales, gills, live in water, cold blooded, lay eggs	Shark, goldfish
	Amphibians	Smooth, moist skin; live in water and on land, cold blooded, lay eggs	Frog, toad
	Reptiles	Scales, lungs, cold blooded, lay eggs	Snake, lizard
	Birds	Feathers, lungs, warm blooded, lay eggs	Cardinal, eagle
	Mammals	Hair or fur, lungs, warm blooded, live birth	Bear, human, mouse

Animal Behavior

Animal **behavior** can be learned or be innate. Innate behaviors are passed from parents to offspring. Innate behaviors are not taught; they help an animal survive or reproduce. We can observe two types of innate behaviors:

1. **Instincts** are complex behaviors in response to a stimulus, and these behaviors are passed down from generation to generation. Two examples are a spider spinning a web and a bird building its nest.
2. **Reflexes** are responses that always occur when a specific stimulus is present; they are automatic. For example, if someone throws a piece of paper at your face, you will blink to protect your eyes.

A well-known example of a reflex is shown in the image below. A strike just below the kneecap will cause the foot to swing forward a bit.

Learned behavior, as its name suggests, is behavior that must be taught. For instance, when the fire alarm rings, students know that they must leave the building, following the appropriate procedures taught in a fire drill.

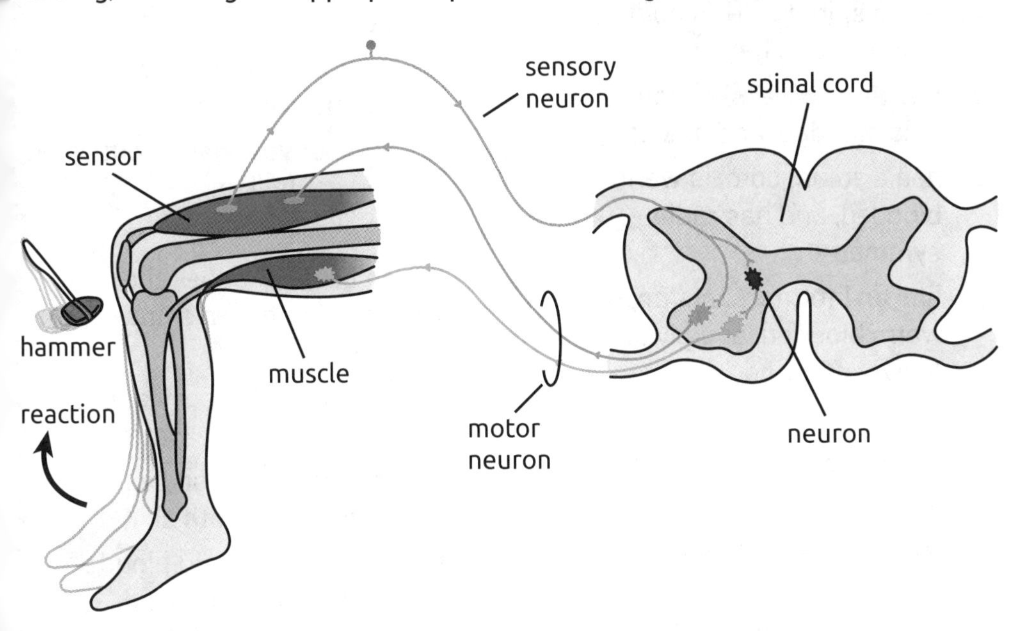

When the knee is struck by a hammer, a sensory neuron carries a message to the spinal cord, and the instructions for the reaction are sent back via a motor neuron, which will cause the muscle to move. This message does not pass through the brain, just the spinal cord.

Lesson Practice

KEY POINT!

Remember that simpler animals generally have simpler structures and less complicated ways of accomplishing life activities.

Complete the activities below to check your understanding of the lesson content.

Skills Practice

Answer the questions based on the content covered in the lesson.

1. A(n) ________ is cold blooded, lays eggs, and lives on land for all of its life.
   A. amphibian
   B. cnidarian
   C. mammal
   D. reptile

2. Which best describes the characteristics of a mammal?
   A. has a spinal cold, is cold blooded, and has hair
   B. has an internal skeleton, is cold blooded, and has lungs
   C. has a spinal cord, is warm blooded, and has radial symmetry
   D. has an internal skeleton, is warm blooded, and gives birth to live offspring

3. Which correctly compares fish and amphibians?
   A. Both are cold blooded, but only fish lay eggs.
   B. Both have scales, but only amphibians lay eggs.
   C. Both are cold blooded, but only fish have scales.
   D. Both have scales, but only amphibians are cold blooded.

4. Which of the following is a reflex?
   A. You tie your shoelaces.
   B. You go to sleep at night.
   C. You laugh at a funny joke.
   D. You move your hand away from a hot stove.

5. Which is the simplest animal?
   A. earthworm
   B. goldfish
   C. oyster
   D. sponge

6. Which of the following is an instinct?
   A. A lizard drinks water off a leaf.
   B. A dog sits when waiting for a treat.
   C. Bees dance to show the location of food.
   D. You put your hands out when you are falling.

7. Which best describes the difference between a snake and a toad?
   A. The snake lives on land; the toad lives in the water.
   B. The snake has scales; the toad has smooth skin.
   C. The snake is cold blooded; the toad is warm blooded.
   D. The snake gives birth to live young; the amphibian lays eggs.

8. Which characteristic do birds and mammals share?
   A. Both have hair.
   B. Both are warm blooded.
   C. Both have an exoskeleton.
   D. Both give birth to live young.

See page 171 for answers and help.

Like all living things, humans need to carry out life activities, which include nutrition, respiration, transport, and regulation. In order to accomplish these activities, the organ systems of the body must work together. **Metabolism** is a term used to describe all the chemical activities that occur in the body.

Movement

The skeletal and muscular systems work together to enable movement. The **skeletal system** consists of your skeleton, which is made up of bones. The **muscular system** is made up of two types of muscles, voluntary and involuntary. **Voluntary muscles** are those that you can voluntarily control, such as the muscles of your arms and legs. These muscles pull bones, which allow you to walk, run, and throw a ball. Other types of movement are also required by the body, movements that you do not even think about, such as the muscles that allow you to breathe and the muscles that move food through your digestive system. These muscles are called smooth or involuntary muscles, because you do not volunteer to move them—they move without your even thinking about it.

Respiration

The **respiratory system** is responsible for gas exchange. The body needs to take in oxygen in order to release energy from the food we eat. The body also needs to release respiration's waste product, which is carbon dioxide.

When you breathe in, or inhale, air moves into your nose (or mouth). It moves down through the trachea, which then branches into your bronchial tubes. The bronchial tubes branch off, becoming smaller and smaller and ending in tiny sacs called alveoli. Capillaries (tiny blood vessels) surround the alveoli, and it is here that gas exchange occurs. Carbon dioxide leaves the capillaries and moves into the alveoli for release through the respiratory system when you exhale. Oxygen moves from the alveoli into the capillaries and from there is carried by the circulatory system to the cells of the body.

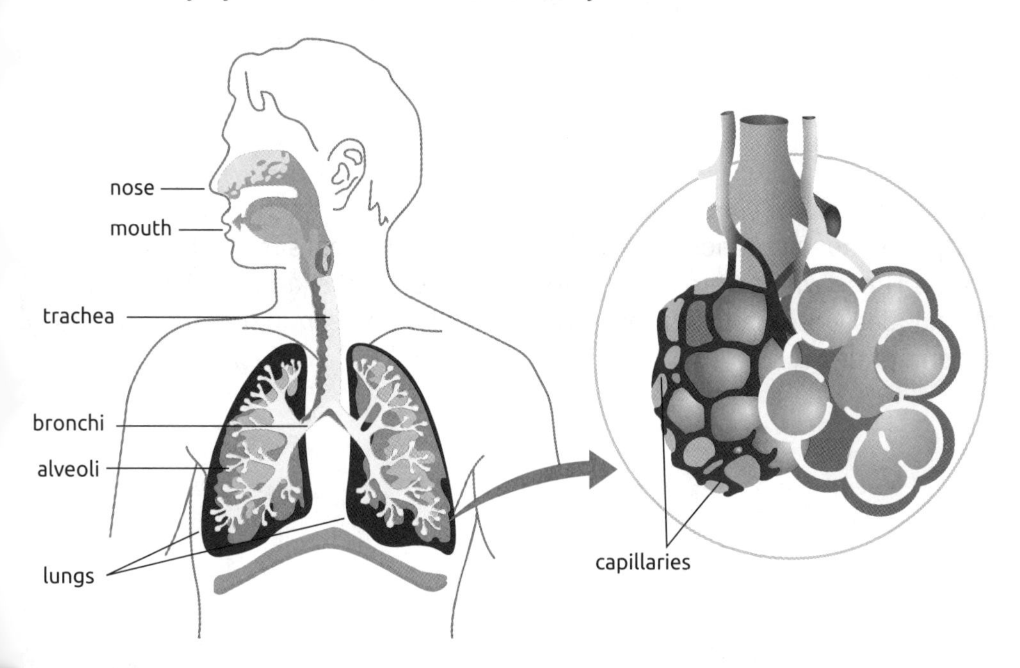

The respiratory system

KEY WORDS

- circulatory system
- digestive system
- endocrine system
- enzymes
- esophagus
- excretory system
- metabolism
- muscular system
- nervous system
- nutrition
- respiratory system
- sense organs
- skeletal system
- urine
- voluntary muscles
- white blood cells

Transport

The organs of the **circulatory system** are responsible for transporting materials around the body. The heart is a muscular organ that controls the movement of blood through the blood vessels. Blood is pushed out of the heart through a large artery (arteries carry blood away from the heart) called the aorta. The aorta splits into smaller and smaller arteries, until the blood vessels become capillaries, which are one cell thick. These blood vessels are where exchange of materials occurs, which is why their walls are so thin—so molecules of materials, such as sugars and gases, can pass easily into and out of them.

The blood is made up of a number of components: **white blood cells**, which help with immune functions, red blood cells, which carry oxygen, and a liquid called plasma, in which the cells are suspended. Capillaries bring materials to the cells of the body and also take them away. For instance, carbon dioxide is brought back from the cells to be exhaled. Capillaries that exchange materials to and from body cells connect to larger vessels that become veins, which carry blood back to the heart.

Nutrition

Nutrition is the process by which living things obtain and process the nutrients needed for survival. The **digestive system** is responsible for breaking down the food you eat into the nutrients required for activities such as growth and repair.

Digestion begins in the mouth, where digestive **enzymes** (chemicals that speed up chemical reactions) are present. Food then moves down the **esophagus** to the stomach. In the stomach, glands (organs that release compounds) release more digestive enzymes to continue the process. Partially digested food then moves into the small intestine, where the most digestion will occur as enzymes are released. Glands of the small intestine complete the digestive process. The nutrients are absorbed through capillaries in the small intestine, where they are then transported by the circulatory system to other parts of the body for use or storage. Any food that is not digested proceeds to the large intestine and is eliminated as waste.

Excretion

The **excretory system** eliminates the waste of the activities that occur in the body. This does not include digestive waste, only waste from chemical reactions. These wastes are filtered from the blood in the kidneys. The waste, which includes water, is then stored in the bladder and eliminated from the body in the form of **urine**.

Regulation

There are two systems whose primary function is to regulate internal body conditions. The **endocrine system** is composed of glands that release hormones, or chemical messengers. For instance, the pancreas is a gland that makes and releases a hormone called insulin, which regulates the amount of sugar in the blood. Several systems are often involved in regulation; in this case, the endocrine system is monitoring a product of digestion and is using the circulatory system to affect that target.

The **nervous system** is made up of the spinal cord, brain, and nerves. The brain can be considered the control center of the body. This organ receives information from the body and environment through the many nerves that run to, from, and within the brain, and then it directs actions. Your body receives information from the outside environment through your **sense organs**, which include your eyes (visual information), nose (smells), skin (touch sensations), ears (audio information), and tongue (tastes). For example, your eyes collect visual information when you see a friend; this information is transported by nerve cells to the brain. Nerve cells, or nerves, transmit impulses to and from the brain and spinal cord. The brain interprets this information—*That is my friend!*—and then sends signals from the brain to the spinal cord to the muscles of your arm to wave at your friend.

Lesson Practice

KEY POINT!

Systems often work together to accomplish life activities. For instance, materials made by one system are transported where they need to go by the circulatory system.

TEST STRATEGY

Treat each answer option as a true/false question. Jot down why each is true or false.

Complete the activities below to check your understanding of the lesson content.

Skills Practice

Answer the questions based on the content covered in the lesson.

1. Nutrients are taken from the small intestine to parts of the body where they are needed. This is an example of which systems working together?
   A. circulatory and digestive
   B. digestive and endocrine
   C. respiratory and circulatory
   D. endocrine and respiratory

2. Someone who has to urinate frequently is most likely having trouble with which system?
   A. circulatory
   B. digestive
   C. excretory
   D. skeletal

3. Where might voluntary muscles be found?
   A. heart
   B. lungs
   C. upper arm
   D. small intestine

4. Which systems keep the conditions in the body at the levels required?
   A. nervous and endocrine
   B. endocrine and excretory
   C. respiratory and nervous
   D. excretory and respiratory

5. Metabolism is a term used to describe all the chemical activities that occur in the body. Which system is responsible for removing the waste products of metabolism?
   A. digestive
   B. excretory
   C. muscular
   D. nervous

6. Which organ interprets information coming in from the senses?
   A. brain
   B. heart
   C. kidney
   D. lung

7. Enzymes work to break down food in which organs?
   A. stomach, mouth, and small intestine
   B. mouth, esophagus, and small intestine
   C. large intestine, stomach, and small intestine
   D. small intestine, large intestine, and esophagus

8. The thyroid gland regulates a person's metabolism. If a person's thyroid is underactive, which may result?
   A. The person may feel cold and be tired.
   B. The person may feel warm and have loose stools.
   C. The person may have an excess of energy and feel warm.
   D. The person may have loose stools and have excess energy.

9. High fevers can be very dangerous because high temperatures can break down the structures of enzymes. Which of the following might occur if a person's enzymes are not functioning properly?
   A. Blood would not be filtered by the kidneys.
   B. Electric messages would not reach the brain.
   C. Food would not be broken down in the stomach.
   D. Chemical messages would not be sent from glands.

10. In order for you to get energy from the food you eat, which system must be functioning properly?
    A. excretory
    B. nervous
    C. respiratory
    D. skeletal

11. Oxygen passes into the muscles of the body through the _______.
    A. alveoli
    B. arteries
    C. capillaries
    D. veins

12. You hear some good news. Which best explains the path of your response in the nervous system?
    A. brain → spinal cord → nerves → muscles → you smile.
    B. ear → nerves → brain → nerves →muscles → you smile.
    C. brain → nerves → ear → nerves → spinal cord → muscles → you smile.
    D. ear → nerves → brain → nerves → spinal cord → nerves → muscles → you smile.

See page 172 for answers and help.

Health

KEY WORDS

- disease
- environmental diseases
- immunity
- immunizations
- immunodeficiencies
- inflammation
- pathogens

The immune system is the body's army—it protects the body from disease. Disease can come from a number of sources, some that are environmental and some that are genetic.

Types of Disease

A **disease** is a condition that negatively affects the body. Infectious diseases are caused by pathogens. **Pathogens** are disease-causing organisms, such as bacteria or viruses. An infectious disease can spread when a person comes in contact with a substance or person contaminated by a pathogen. Infectious diseases include colds and the flu.

Noninfectious diseases are caused by other factors, including heredity, age, and chemicals in the environment. As the name states, the environment causes **environmental diseases**. Mercury and lead poisoning are conditions that one can develop due to exposure to harmful chemicals in the environment.

Hereditary diseases are conditions caused by factors that have been passed on from parents to offspring; often this is due to a mutation or change in the genes. Down syndrome is an example of a genetic condition.

Age-related diseases are illnesses and conditions that generally affect older people. These are conditions that arise as a result of years of use. For instance, osteoarthritis is caused by wear and tear on the joints and can result in pain and swelling.

Types of Immunity

Immunity is the body's ability to fight off a disease. Everyone is born with a natural (or innate) immunity to certain pathogens. This includes the first line of defense against pathogens, such as the skin, mucus (which traps bacteria and small particles), and the reflex that makes you cough.

Acquired (or adaptive) immunity develops during one's lifetime. Acquired immunity is due to a buildup of immune cells in response to diseases the body is exposed to. This explains why children are often sick more frequently than adults—adults have had more time to develop immunity to more diseases.

Acquired immunity is also strengthened by **immunizations** (or vaccinations) that you receive. When you receive a vaccination, you are getting an injection of a weakened or dead pathogen, so it cannot make you sick. The body recognizes the pathogen as foreign and builds up defenses to it so that the next time the body is exposed, it will be ready and can easily defeat the pathogen.

Immunodeficiencies are caused by an immune system that is not working properly. Some people are born with immunodeficiencies, while other immunodeficiencies are the result of drugs or infections.

The Immune Response

Once a pathogen makes it past the first line of defense and enters the body, a person is infected. At this point, the immune system gets to work. When tissues are damaged by a pathogen or a trauma, the injured cells release chemicals that cause blood vessels to leak fluid into the area, which causes swelling, or **inflammation**. The inflammation allows the area to be blocked off and also attracts white blood cells that will fight off the pathogen.

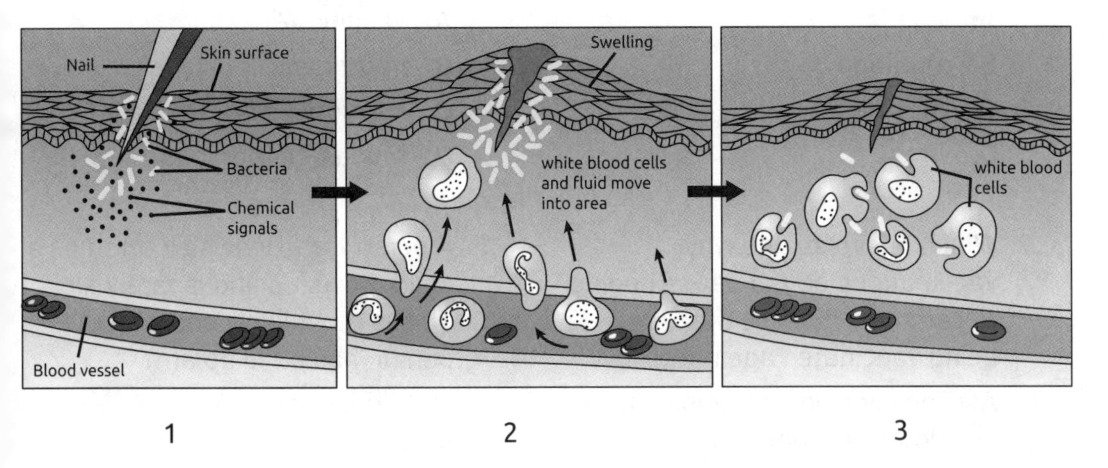

1. Nail punctures skin, allowing pathogens to enter the body. Chemicals released by damaged tissue relay message of tissue injury.

2. Blood vessels leak fluid into area, and swelling occurs. White blood cells move into area where infection has occurred.

3. White blood cells destroy bacteria and debris, and healing begins.

Interactions between the Immune System and Other Body Systems

In order to function properly, the immune system coordinates with other body systems. For instance, the circulatory system transports the immune cells around the body. The blood cells involved with immune function grow and mature within the bone marrow of the skeletal system. The respiratory system also provides a line of defense for the immune system because it is covered in mucus, which can trap foreign substances.

Lesson Practice

KEY POINT!

We acquire immunity because white blood cells build up a memory to individual pathogens.

TEST STRATEGY

Underline the key words and phrases in the question and passage, table, graph, or diagram. For items with a passage, underline the key words and phrases in the question, and then look for them in the passage.

Complete the activities below to check your understanding of the lesson content.

Skills Practice

Answer the questions based on the content covered in the lesson.

1. Which of the following is an infectious pathogen?
   A. lead
   B. mutation
   C. vaccine
   D. virus

2. Which best explains why vaccinations do not cause people to catch the disease they are being vaccinated against?
   A. The vaccine contains white blood cells only.
   B. The vaccine contains dead or weakened pathogens.
   C. The vaccine contains environmental blockers of the disease.
   D. The vaccine contains medicine that will help cure the disease.

3. If a mother has a condition that her mother had, and she passes the condition along to her child, it is a(n) _______ disease.
   A. age-related
   B. environmental
   C. hereditary
   D. infectious

4. Which of the following is part of a person's innate immunity?
   A. vaccines
   B. pathogens
   C. memory cells
   D. mucus membranes

5. A person who was sick often as a child grows up and is rarely sick. This is most likely because the person has build up a(n) _______ immunity.
   A. acquired
   B. inborn
   C. innate
   D. natural

6. Why is inflammation important?
   A. It helps prevent an infection from spreading.
   B. It is the first signal that a trauma has occurred.
   C. It shows that the pathogen is no longer infectious.
   D. It weakens blood vessels so white blood cells can get to the area.

7. What might happen if a person has an immunodeficiency?
   A. He cannot fight infections.
   B. He fights infections more quickly.
   C. He does not develop inflammation.
   D. He develops inflammation more easily.

8. Why is coughing a form of natural immunity?
   A. Coughing helps attract immune cells.
   B. Coughing helps the inflammation process.
   C. Coughing allows white blood cells to reach the infection.
   D. Coughing prevents pathogens from getting into the lungs.

See page 172 for answers and help.

Organisms

KEY WORDS

- cladogram
- domains
- eukaryotes
- kingdoms
- prokaryotes
- species
- taxonomy

Life on Earth is complex and diverse. Scientists have identified and named about 1.78 million different types of organisms, or species, but it is estimated that as many as 30 million species may be on Earth. Scientists make sense of all of this diversity by naming, sorting, and classifying organisms into groups.

Classification of Organisms

Taxonomy is the naming and classification of living organisms. Scientists classify living things based on similarities and differences in their (1) internal and external structure, (2) chemical makeup, and (3) evolutionary relationships. Organisms are placed into groups with similar, closely related organisms. Those groups are combined to make larger groups, with less similar organisms.

The largest classification groups are the three **domains** of life: Eukarya, Bacteria, and Archaea. Eukarya includes all of the **eukaryotes**, organisms that have cells with their DNA contained in a membrane-bound nucleus. Organisms in the Bacteria and Archaea domains are **prokaryotes**, which have cells without organelles or nuclei. Prokaryotes are single-celled, or unicellular, organisms, while eukaryotes can be unicellular or multicellular. In general, prokaryotes are simpler organisms, and eukaryotes are more complex organisms.

The domains are divided into smaller categories called **kingdoms**, which are further divided into smaller groupings. From largest to smallest, the major taxonomic ranks are domain, kingdom, phylum, class, order, family, genus, and species. The species is the most basic unit of classification for organisms. A **species** can be defined as a group of individuals that can potentially interbreed in nature. Their offspring must also be able to interbreed.

The following table shows the complete taxonomy for an example species, the lion.

Taxonomic Rank	Classification
Domain	Eukaryota (eukaryotes)
Kingdom	Animalia (animals)
Phylum	Chordata (animals with a spinal cord)
Class	Mammalia (mammals)
Order	Carnivora (meat eaters)
Family	Felidae (cats)
Genus	*Panthera* (big cats)
Species	*leo* (lion)

The Scientific Naming System

The same organism can have many different names in different places and languages. Different organisms can also have the same "common name." For example, there are three unrelated birds, all called "robins," in Europe, North America, and Australia. To avoid confusion, scientists use standardized Latin names for species. The scientific name for a species is a two-word Latin phrase made up of its genus and species names. For example, the scientific name for lion is *Panthera leo*, since the lion's genus is *Panthera* and its species name is *leo*.

Cladograms

In the past, scientists grouped organisms based on the way they looked or behaved. Now, scientists can compare species' DNA to determine how closely related they are. Modern taxonomists try to organize living things based on their evolutionary relationships, so that all of the species that appear to have evolved from a common ancestor fall into the same taxonomic group.

The relationships among species and larger groups can be shown in a **cladogram**. Cladograms are branching diagrams, like family trees. In these diagrams, each branching point shows where scientists believe that two species diverged from a common ancestor. Species closer together in a cladogram share characteristics that are not shared with species that branched off earlier. Species with fewer branching events separating them are believed to be more closely related evolutionarily.

This simplified cladogram shows relationships among some vertebrate groups. A complete tree would have many more branches.

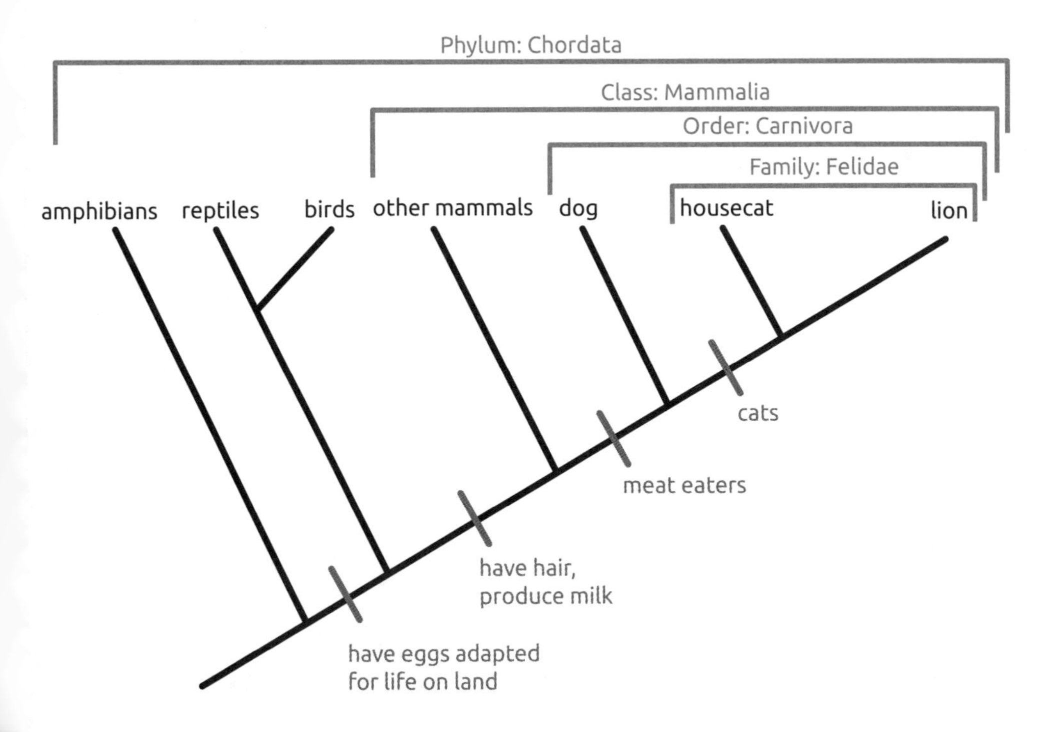

Cladogram showing selected vertebrates

Lesson Practice

KEY POINT!

Taxonomy is always changing, as scientists discover new information about the relationships among species.

TEST STRATEGY

Look for any familiar terms and think about or jot down their meanings. For any unfamiliar terms, try to break down the words to determine meaning, using prefixes, suffixes, stems, or similar-sounding words that you are familiar with.

Complete the activities below to check your understanding of the lesson content.

Skills Practice

Fill in the blanks in the cladogram below with the four species given, writing the letter A, B, C, or D on the blanks. The final diagram should reflect the evolutionary relationships among the species.

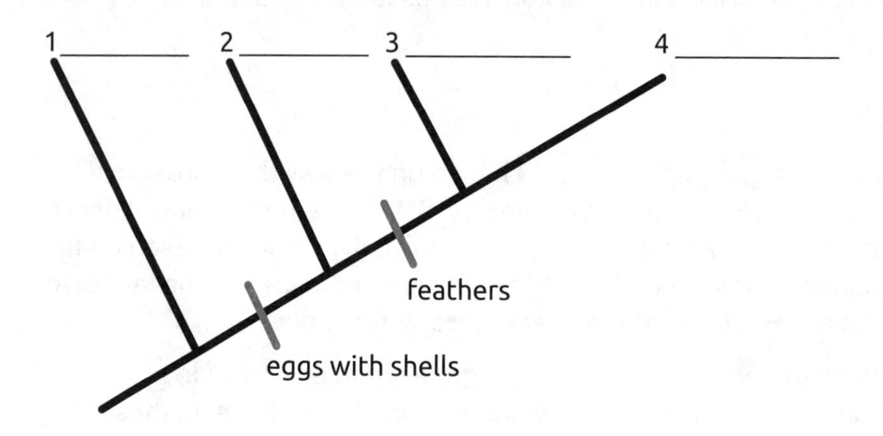

A. crocodile

B. crow

C. dog

D. penguin

The following table shows the taxonomy of human beings. Use it to answer questions 5–10.

Fill in the missing taxonomic ranks in the left column.

Taxonomic Rank	Example
5.	Eukaryota (eukaryotes)
Kingdom	Animalia (animals)
6.	Chordata (animals with a spinal cord)
Class	Mammalia (mammals)
7.	Primates (primates)
8.	Hominidae (great apes)
9.	*Homo* (hominins)
Species	*sapiens* (modern humans)

10. What is the two-word scientific name for modern humans?

____

11. What is the most important feature that modern taxonomists consider when classifying organisms into groups?
    A external structural features (like wings or fur)
    B behavior
    C genetic relatedness
    D internal structures (like digestive systems)

12. What is a species? Write a definition in your own words.

____

____

See page 172 for answers and help.

Heredity

KEY WORDS

- alleles
- chromosomes
- DNA
- genes
- genetics
- genotype
- heredity
- mutation
- phenotype

Throughout history, people have noticed that organisms can pass on traits to their offspring. For example, a child with tall parents is more likely to be tall than a child with short parents. This is called **heredity**. In recent decades, scientists have made great strides in understanding how heredity works.

DNA and Genes

All of the information required for an organism to develop and function is stored in that organism's genetic material, called **DNA**. In eukaryotes, DNA is organized into **chromosomes**, threadlike structures made of proteins packaged with DNA. Humans have 23 pairs of chromosomes, but other organisms may have many more or less. Chromosomes contain **genes**, sections of DNA that determine particular characteristics of the organism by providing instructions for building proteins. Each gene has a specific location on the chromosome. Different versions of the same gene are called **alleles**. For example, in humans, an eye color gene could have an allele for brown eyes and an allele for blue eyes. **Genetics** is the study of how genes work and how they are transferred from parents to offspring.

Most eukaryotic cells contain two copies of each chromosome. For example, every human cell has two copies of 23 chromosomes, for a total of 46 chromosomes. In species that reproduce sexually, like most animals and plants, an individual gets one copy of each chromosome from each of its parents.

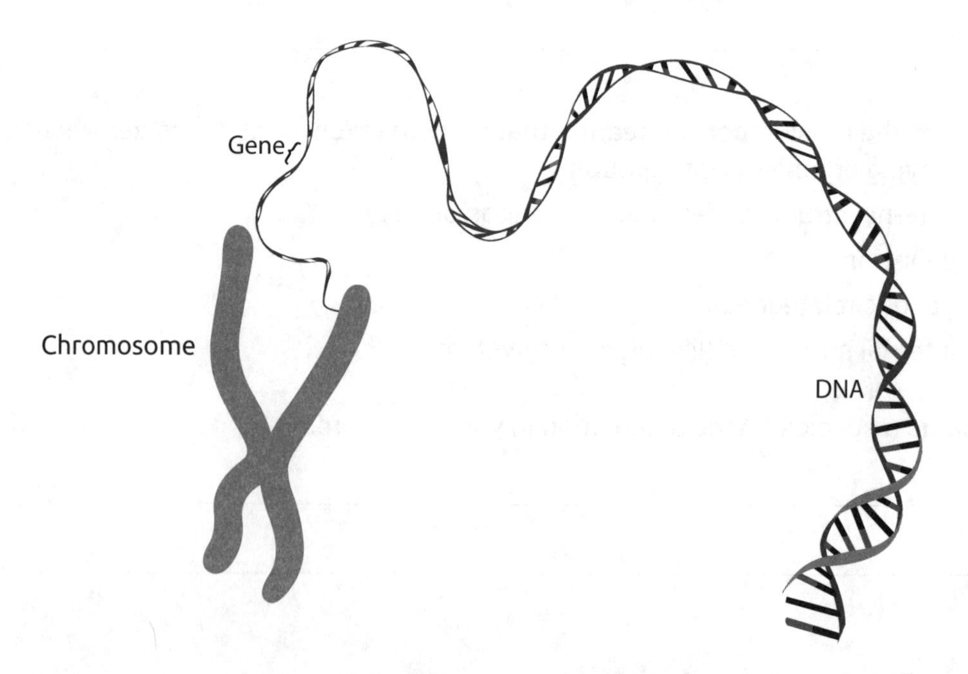

Genes are found at specific locations on the chromosome.

Mutations and Genetic Disorders

When DNA is copied to make new cells, sometimes a mistake occurs, producing a copy that is slightly different from the original molecule. This is called a **mutation**. If a mutation occurs in a gene, the new gene may be different from the original, or it may not work at all. Mutations that are passed down from parents to offspring can lead to new traits. These traits may be useful, harmful, or have no impact on the offspring's chances of survival. Genetic disorders are diseases caused by genes that do not function normally. For example, sickle cell anemia is a disease caused by a mutation in the gene that provides the instructions for making the protein that carries oxygen in the blood.

Gene Expression and Dominance

In sexual reproduction, each parent gives one copy of each chromosome to its offspring. This means that a child has two copies of each gene: one from its father, and one from its mother. The two parents can provide different alleles of the same gene, or both parents can provide the same allele. The specific alleles of a gene in an organism is called its **genotype**. The trait that is expressed in an organism is called its **phenotype**. For example, the tendency to have dimples is inherited. A child could get one gene from her father with the allele for dimples, and one gene from her mother with the allele for no dimples. Her genotype would be one "dimples" allele and one "no dimples" allele. In this case, the child would most likely have dimples. Her phenotype would be dimples.

In cases like this, when organisms with two different alleles of a gene have the trait associated with one of the alleles, that allele is said to be dominant. The allele for the trait that is not expressed is called recessive. The allele for dimples is dominant over the allele for no dimples. If an allele is truly dominant, an organism that has that allele will always exhibit the phenotype associated with that allele, even if it also has a different (recessive) allele. In reality, most traits are governed by more than one gene. This means that the trait expressed is not always the one that matches the dominant allele of a gene.

KEY POINT!

Remember that traits you can observe, like eye color, represent phenotypes, rather than genotypes.

Punnett Squares

A Punnett square is a diagram that can be used to predict what genotypes are possible for offspring from parents with known genotypes. It can also predict the probability the genotypes will occur, or how many offspring are likely to have each genotype. In a Punnett square, the two alleles from one parent are written at the top, and the two alleles from the other parent go on the left side. Each box within the square represents a possible combination of alleles from the two parents. The distribution of genotypes in the square is the same as the expected distribution among the offspring of the parents. If half of the boxes contain a given genotype, the offspring have a 50% chance of inheriting that genotype.

In the example shown here, the allele for dimples is shown as a *D*, and the allele for no dimples is shown as a *d*. The Punnett square shows the possible genotypes for children of a mother with both alleles and a father with only the no dimple allele.

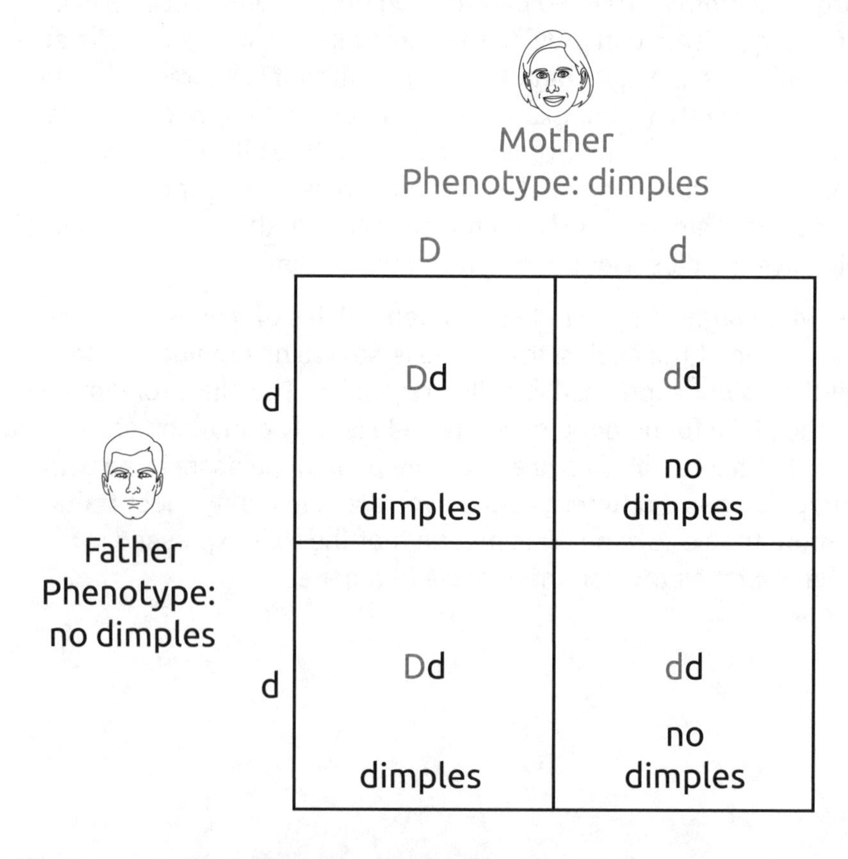

What percent of the children are likely to have dimples?

Complete the activities below to check your understanding of the lesson content.

Skills Practice

Fill in the blank in each sentence with a word from the word bank.

alleles	chromosomes	DNA
genotype	mutations	phenotype

1. Chromosomes are made of proteins and ____________________.

2. Most human cells have 23 pairs of ____________________.

3. New versions of genes can be created by ____________________, which can be harmful or helpful.

4. A genetic trait expressed by an organism is called a(n) ____________________.

5. A girl has blue eyes, and her sister has brown eyes. The sisters have different ____________________ for eye color.

6. The alleles of a particular gene that an organism has inherited define its ____________________.

Answer the following questions based on the content covered in the lesson.

7. A botanist is breeding a species of orchids that can have either pink or white flowers. She finds that when both parents are pink, most of the offspring are pink, but some are white. When both parents are white, all of the offspring are white. Which allele of the flower color gene is likely dominant?

____

TEST STRATEGY

Rewrite the question in your own words to make sure you understand it.

Lesson Practice

8. A man named Joe has had many experiences in his life. Which of these experiences might produce a change in Joe that he could pass down to his children?

    A Joe runs 10 miles every day and his lungs become stronger.

    B Joe is exposed to air pollution that causes a mutation in his DNA.

    C Joe learns five languages.

    D Joe has an accident that leaves him with a scar on his arm.

9. Identical twins have identical DNA, while fraternal twins do not. Imagine that a set of identical twins and a set of fraternal twins all have brown eyes. Which of the following is NOT true about the gene for eye color in these four people?

    A The identical twins must have the same genotype.

    B The fraternal twins must have the same phenotype.

    C All four must have the same genotype.

    D All four must have the same phenotype.

10. The allele for dimples is dominant, and the allele for no dimples is recessive. If most people don't have dimples, what does this tell us about how common the two alleles are in the population?

    A The allele for no dimples is much more common than the allele for dimples.

    B The allele for dimples is much more common than the allele for no dimples.

    C The two alleles are equally common in the population.

    D The frequency of the two alleles in the population is unrelated to the number of people who actually have dimples.

The following Punnett square shows the alleles for blood type for two parents. The A allele, which produces type A blood, is dominant, and the O allele, which produces type O blood, is recessive. Use the square to answer questions 11–17.

		Mother	
		A	O
Father	O	11	12
	O	13	14

Fill in the missing genotypes from the Punnett square. Write your answers on the lines below.

11. ____________________

12. ____________________

13. ____________________

14. ____________________

15. What fraction of these parents' children would you expect to have the same genotype as their mother?

____

16. What is the phenotype of the father?

____

17. What fraction of the children would you expect to have type O blood?

____

See page 173 for answers and help.

Evolution

KEY WORDS

- adaptation
- common ancestor
- diversity
- evolution
- fossil record
- natural selection
- species
- traits

Two years after Charles Darwin published *On the Origin of Species*, a discovery was made in Southern Germany to further validate his ideas. A fossil about 150 million years old was uncovered. Like the dinosaurs of the past, it had jaws with teeth and a long bony tail. However, it also had broad wings and feathers, like the birds we see today. This fossil, called *Archaeopteryx* was the "missing link" between birds and dinosaurs. Because of this discovery and other evidence, scientists believe that the dinosaurs were the ancestors of birds.

The Fossil Record

Remains of plants and animals, known as fossils, can be found in sedimentary rock deposits all over the globe. When sedimentary rock layers are formed, the newer layer is deposited over the older layer. This allows scientists to use fossils as a record of life over the course of Earth's history, called the **fossil record**. The fossil record not only shows that life has existed, but also that life has changed over large periods of time. Various life forms that are part of the same group are known as **species**. Some **traits**, or features, of extinct species are found in transitional species between organisms. For example, *archaeopteryx* had traits of both dinosaurs and modern birds.

***Archaeopteryx* has traits of both dinosaurs and modern birds.**

Evolution

The fossil record shows scientists that **evolution** has occurred, or that genetic change has happened over time. At the basis of evolution is the idea that all species that have ever lived share a **common ancestor**. From this common ancestor, many different species have branched off, giving life the **diversity** we know exists today. The trees in the park, your pet cat, and the fish in the river all came from the same ancestor.

KEY POINT!

Evolution occurs because of descent with modification. Genetic changes happen from one generation to the next.

How Evolution Occurs

In 1859, the naturalist Charles Darwin published *On the Origin of Species*. Earlier in his life, he had traveled on the HMS Beagle to South America, and during his journey, he was struck by the variation he observed among and between species. One of his most famous observations was about the ground finches in the Galapagos Islands. He noted that among the finches, their beak shape varied according to the type of food they ate. This led Darwin to wonder about how diversity in life occurs.

Darwin's theory of **natural selection** explained that nature selects traits that are advantageous to the species. Take, for example, a population of beetles living on a tree. In this population, some are brown, and some are green. After a forest fire, the green leaves are burned away, and the green beetles stand out and are easily picked off and eaten by birds. Therefore, far fewer green beetles are left to survive and reproduce than brown beetles. Brown beetles reproduce with other brown beetles and produce brown-colored offspring, and after several generations, the entire population of beetles is brown.

The brown coloring of the beetles in this scenario is considered an **adaptation**—a common trait in a population that provides an improved survival function. Natural selection occurs because of variation within a species; if all the beetles were green from the start, then the entire population would have been killed off.

Lesson Practice

TEST STRATEGY

Cover the answer choices. Read the question and come up with an answer on your own. Then, choose the answer that best matches your own.

Complete the activities below to check your understanding of the lesson content.

Vocabulary

Write definitions in your own words for each of the key terms.

1. adaptation ____________________

____

2. natural selection ____________________

____

3. common ancestor ____________________

____

Skills Practice

Answer the questions based on the content covered in the lesson.

4. Why is variation within a species necessary for evolution to occur?

____

____

5. The katydid is an insect that has a leaf-like appearance. How did this adaptation most likely come about?

A The common ancestor of the katydid had a leaf-like appearance.

B When an insect population found leaves to be a suitable habitat, they changed their shape.

C Some members of an insect population were shaped more like leaves than others, and they lived to reproduce.

D Katydids who lived on leaves had genes that were more easily passed to the next generation than those who lived in other areas.

See page 173 for answers and help.

Surrounding us are living and nonliving things: from the air we breathe, to the trees outside our windows, to the soil beneath our feet, to the birds flying overhead. Although we may not always be aware of these components that make our immediate environment, they are part of our ecosystem.

KEY WORDS

- abiotic components
- biome
- biotic components
- ecosystem
- minerals

What Is an Ecosystem?

An **ecosystem** consists of all the living and non-living components in a particular environment. For example, in a forest ecosystem, the **biotic** (living) **components** can include trees, birds, deer, rabbits, wolves, and shrubs. In the same ecosystem, the **abiotic** (non-living) **components** can include the gases in the air, the soil beneath the plants, the light from the sun, and the rocks and other **minerals** in the ground.

It is possible for members of one ecosystem to interact with members of other ecosystems. However, they are much more likely to interact with members of their own ecosystem.

Communities of the World

A **biome** is a community existing of many ecosystems. These biomes are classified according to different factors. These include the climate of the area, the main vegetation of the area, and adaptations of the species living in that particular area.

Some scientists divide the world into six biomes. These include forest, grassland, freshwater aquatic, marine aquatic, desert, and tundra. Others use a more specific method of classification to divide up biomes. For example, forests can be further classified as tropical rainforests, temperate deciduous forests, and taiga forests.

The boundaries between biomes are not rigid; in fact, they are far from it. In some areas, transition zones exist between biomes—for example, coasts and wetlands are found between land and aquatic biomes. In addition, biomes move with changes in climate; for instance, ten thousand years ago, part of the Sahara desert was occupied by lush vegetation.

KEY POINT!

Ecosystems consist of living and non-living components that interact with each other.

Nutrient Cycling

Some nutrients, including carbon and nitrogen, are transferred through an ecosystem. This nutrient cycling is accomplished through processes such as uptake, respiration, and decomposition. Both biotic and abiotic components play key roles in nutrient cycling.

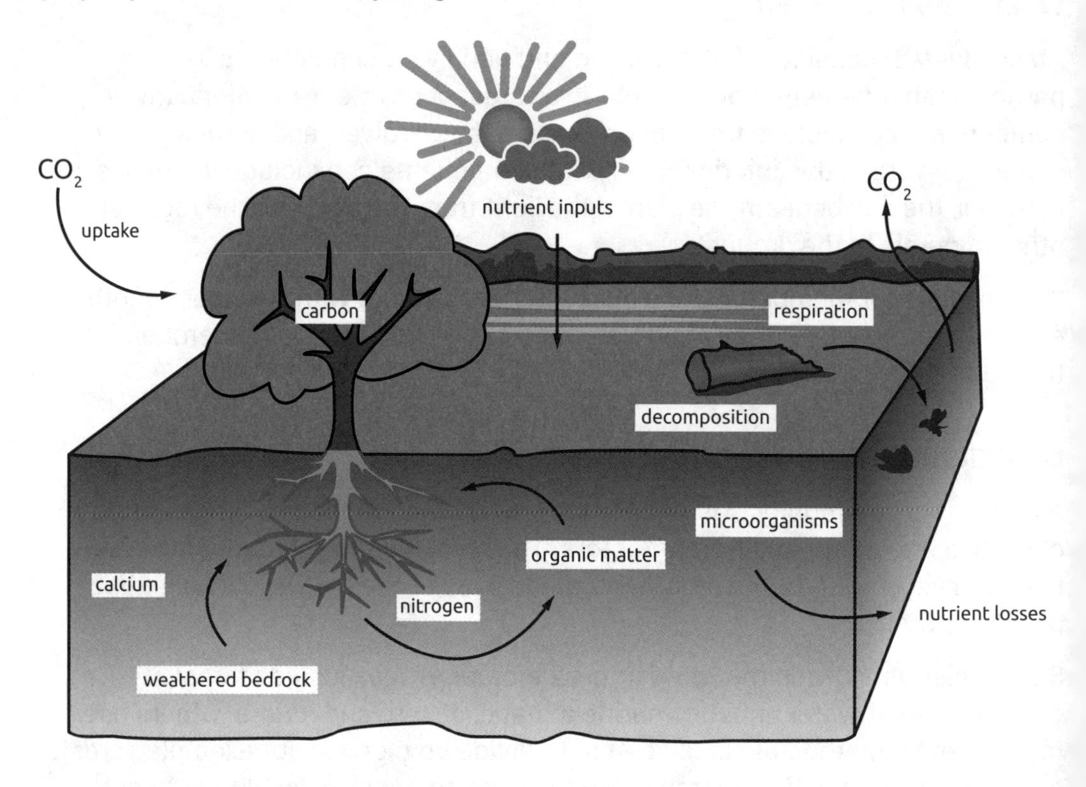

Many different pathways are involved in nutrient cycling. Some of those pathways are shown in this diagram.

Complete the activities below to check your understanding of the lesson content.

Skills Practice

Identify whether each of the following components is biotic or abiotic.

1. lion ____________________

2. grass ____________________

3. carbon ____________________

4. oxygen ____________________

Answer the questions based on the content covered in the lesson.

5. In a forest ecosystem, trees are cut down, and a factory is built on the land. Explain how this can affect the ecosystem.

____

____

____

____

6. How is a biome different from an ecosystem?

   A A biome consists of biotic and abiotic components, while an ecosystem does not.

   B A biome has components that are interconnected, while the components of an ecosystem exist independently.

   C A biome has clear boundaries, defined by the vegetation of an area, while an ecosystem has less sharply defined boundaries with transition zones.

   D A biome is a community of living things classified by the climate of the region, while an ecosystem is the interaction of living and non-living things in an environment.

See page 173 for answers and help.

TEST STRATEGY

Draw a sketch describing an ecosystem and a biome, boxes to compare sizes, a web interconnecting key phrases, or any other diagram that represents the information visually.

Food Webs and Symbiotic Relationships

KEY WORDS

- decomposers
- food chain
- food web
- nutrients
- symbiosis
- trophic levels

It is 12:30 pm, and you ate breakfast a long time ago; you are starting to feel sluggish, and your stomach is beginning to rumble. You are aware you must eat something to regain energy—perhaps a burger, or maybe a salad and fresh fruit. Following your meal, you begin to feel recharged and ready to tackle the task at hand. How does food provide energy for your body, and where does that energy come from?

Food Chains

Every living organism on Earth, from the smallest one-celled algae to the blue whale, needs food to survive. A **food chain** describes which organisms are consumed directly by others in an ecosystem. For example, consider these animals living in a forest:

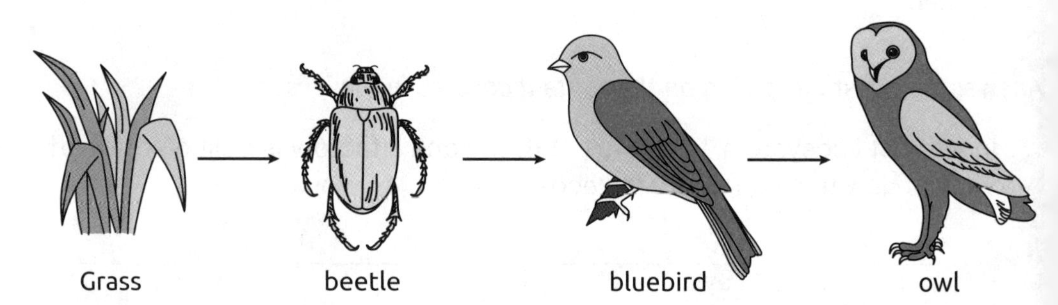

In this food chain, the beetle eats the grass, the bluebird eats the beetle, and the owl eats the bluebird.

The organisms in a food chain are classified into **trophic levels**, or the position the organism occupies in the food chain. Generally speaking, a food chain contains four trophic levels: producers, primary consumers, secondary consumers, and tertiary consumers. Primary consumers eat producers. The other levels of consumers eat other consumers. Some consumers, called carnivores, eat only other animals. Herbivores only consume plants, or producers. Omnivores are consumers that eat both plants and animals.

Decomposers, such as fungi and bacteria, are also part of food chains. They break down decaying organic matter and turn it into inorganic matter, or **nutrients** that are put back into the soil.

Solar energy provides the energy needed for photosynthesis to occur in producers. Photosynthesis allows plants to make glucose, which stores energy that is used to carry out life processes, and this energy is passed up the food chain. However, energy is lost as it is passed from one trophic level to the next.

Food Webs

An organism in a food chain can be eaten by organisms that are not included in that particular food chain. For example, in the forest food chain demonstrated earlier, a bluebird may also be eaten by a snake or a raccoon. A **food web** shows the interconnections of all the species in an ecosystem.

This diagram shows the relationships between typical organisms. The arrows connect the prey (diet) to the predator (consumer).

KEY POINT!

A food chain shows only a small portion of a food web.

Symbiosis

Symbiosis describes the relationship between two different species that have a close physical interaction with one another. This association may last for the lifetime of one or both members of the relationship. We can observe three types of symbiotic relationships:

Type of Symbiosis	Description of Interaction	Examples
Mutualism	Both partners benefit	Bee and flower: bee pollinates flower; bee receives nectar
Commensalism	One partner benefits, and the other is unharmed	Anemones and clownfish: the clownfish is protected from predators by the stinging tentacles of the anemone; the anemone is unaffected
Parasitism	One partner benefits, and the other is harmed	Dog and fleas: fleas get food and a warm home; dog gets bitten and itchy

Lesson Practice

TEST STRATEGY

When you are asked about a general concept on the test, think of a specific example of that concept. This can help you to put the question in perspective and answer it correctly.

Complete the activities below to check your understanding of the lesson content.

Skills Practice

Answer the questions based on the content covered in the lesson.

1. Explain how an organism can be both a primary and secondary consumer.

____

____

2. How can an organism be part of more than one food chain?
   - **A** An organism may die before it can be eaten by a consumer.
   - **B** An organism may be eaten by more than one type of animal.
   - **C** An organism may make its own food at certain points in its life cycle.
   - **D** An organism may lose energy as it moves from one trophic level to the next.

Use the following aquatic food chain to answer questions 3 and 4.

algae → zooplankton → fish → heron

3. Which organism in the food chain is a secondary consumer?
   - **A** algae
   - **B** heron
   - **C** fish
   - **D** zooplankton

4. Which organism in the food chain has the LEAST amount of energy?
   - **A** algae
   - **B** heron
   - **C** fish
   - **D** zooplankton

See page 173 for answers and help.

Answer the questions based on the content covered in this unit.

The following Punnett square shows the alleles for flower color for two parent plants. The B allele, which produces a blue flower, is dominant, and the b allele, which produces white flowers, is recessive. Use the square to answer the first two questions.

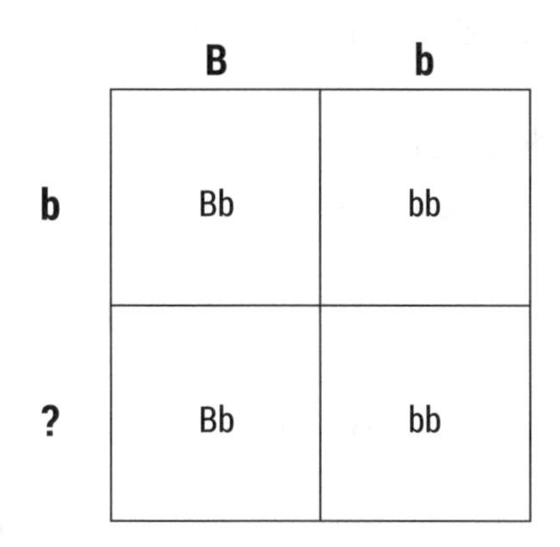

	B	b
b	Bb	bb
?	Bb	bb

1. What is the genotype for the missing parent?
   - **A** BB
   - **B** Bb
   - **C** bb
   - **D** Either Bb or bb

2. What percent of the offspring are expected to be white?
   - **A** 100%
   - **B** 50%
   - **C** 25%
   - **D** 0%

3. The Lopez family has three sons. The older two are identical twins. All three sons have dimples. Which of the following statements about the gene for dimples is NOT true?
   - **A** The twins have the same genotype as each other.
   - **B** The twins have the same phenotype as each other.
   - **C** The younger son has the same phenotype as the twins.
   - **D** The younger son has the same genotype as the twins, but a different phenotype.

4. Which of the following taxonomic ranks is contained within an order?
   - **A** genus
   - **B** class
   - **C** domain
   - **D** phylum

5. A taxonomist has proposed that species A should be in the same family as species B but that species C should be in a different family. Which of the following is the best evidence to support this proposal?
   - **A** Species A and species B both have wings, but species C does not.
   - **B** Species B and species C are the same color, and species A is a different color.
   - **C** DNA evidence suggests that species A and species B share a common ancestor more recently than species C.
   - **D** DNA evidence suggests that species A diverged from the common ancestor of species B and C before species B and C split from each other.

6. Which of the following structures is NOT found in an animal cell?
   - **A** mitochondrion
   - **B** nucleus
   - **C** cell wall
   - **D** endoplasmic reticulum

7. Which of the following features is present in eukaryotic cells but not in prokaryotic cells?
   - **A** cell membrane
   - **B** cell wall
   - **C** DNA
   - **D** membrane-bound nucleus

8. Single-celled organisms, like the cells of our body, obtain needed water through
   - **A** absorption.
   - **B** drinking.
   - **C** germination.
   - **D** osmosis.

9. If the phloem within a plant's stem were severed,
   - **A** the plant would be unable to carry sugar to its roots.
   - **B** the plant would be unable to transport water to its leaves.
   - **C** the plant would be unable to absorb minerals from its roots.
   - **D** the plant would be unable to move carbon dioxide to its leaves.

10. Which best describes why plants do not survive if too many of their roots are cut?
    - **A** They cannot transport water to the roots for storage.
    - **B** They cannot transport sugar to the leaves for storage.
    - **C** They cannot transport sugar to the roots for photosynthesis.
    - **D** They cannot transport water to the leaves for photosynthesis.

11. A mutation in a person's DNA can be especially bad because
    - **A** a protein may not be built properly.
    - **B** an amino acid may not bond properly.
    - **C** the mRNA will not be able to leave the nucleus.
    - **D** the tRNA will not be able to move to the ribosome.

12. Which of the following is NOT a product of cellular respiration?
    - **A** water
    - **B** carbon dioxide
    - **C** ATP
    - **D** glucose

13. Why are plants sometimes referred to as "carbon sinks?"
    - **A** They are made up of primarily carbon.
    - **B** They remove carbon dioxide from the atmosphere.
    - **C** They take up carbon in the soil through their roots.
    - **D** They provide carbon dioxide for cellular respiration.

14. Which of the following is segmented and has an exoskeleton?
    - **A** ant
    - **B** clam
    - **C** earthworm
    - **D** shark

15. Which of the following is an instinctual behavior?
    - **A** A cat will come to its owner when it is called.
    - **B** A baby will grab an object that has been placed in her hand.
    - **C** A dog goes to his feed bowl when he hears the food being poured.
    - **D** After hatching, baby sea turtles wait until dark to venture out into the ocean.

16. An organism has no developed organ systems, but it does have cells that allow it to capture prey. It is a(n)
    - **A** earthworm.
    - **B** jellyfish.
    - **C** squid.
    - **D** sponge.

Answer questions 17–19 based on the following image.

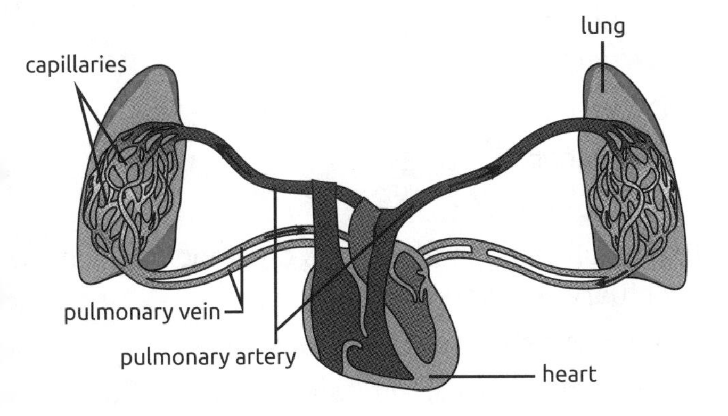

17. Which two systems are working together in the diagram?
    - **A** digestive and excretory
    - **B** excretory and digestive
    - **C** respiratory and digestive
    - **D** circulatory and respiratory

18. Which best describes the path blood takes in the diagram?
    - **A** heart → pulmonary vein → lungs to pick up oxygen → body → pulmonary vein → heart
    - **B** heart → pulmonary artery → lungs to pick up oxygen → pulmonary vein → heart → body
    - **C** heart → pulmonary vein → lungs to pick up oxygen → heart → pulmonary artery → body → lungs
    - **D** heart → pulmonary artery → lungs to pick up oxygen → heart → body → pulmonary vein → lungs

19. Which system controls the functioning of the systems in the diagram?
    - **A** endocrine system
    - **B** excretory system
    - **C** circulatory system
    - **D** nervous system

20. A person's arteries may build up with plaques over time that affect the person's blood flow. This is an example of a(n) _______ disease.
    - **A** age-related
    - **B** environmental
    - **C** hereditary
    - **D** infectious

21. Pus is often seen at the site of an infection. Which best explains the makeup of pus?
    - **A** mucus, living bacteria, and debris
    - **B** mucus, dead bacteria, and pathogens
    - **C** white blood cells, dead bacteria, and debris
    - **D** white blood cells, living bacteria, and mucus

22. Chicken pox is caused by a virus. Why is it important to get the chicken pox vaccine?
    - **A** It prevents you from becoming infected with chicken pox.
    - **B** It inhibits the people around you from getting the disease.
    - **C** It allows white blood cells to fight off all viral diseases.
    - **D** It ensures that inflammation will be the only sign of the disease.

23. Which of the following scenarios describes natural selection?
    - **A** A population of small rodents can no longer find shelter after a forest fire. They move to a new area.
    - **B** A population of mammals stretches their necks to reach leaves on tall trees. Eventually, their necks grow longer.
    - **C** A population of insects wants to be able to better escape predators. They change their color to better blend in with their surroundings.
    - **D** A population of birds has individuals with different-sized beaks. The birds with the smaller beaks can find more food. Over time, the entire population of birds has small beaks.

Review the following passage and diagram. Then answer questions 24–26.

An estuary is a coastal area where freshwater from rivers and streams and saltwater from the oceans meet and mix. This makes an estuary an ecosystem that is habitable to many different organisms.

24. Which of the following components of the estuary ecosystem are abiotic?
    - **A** fish, birds, algae
    - **B** snake, air, algae
    - **C** grasses, water, sediment
    - **D** air, water, sediment

25. Which organism in the estuary ecosystem is a producer?
    - **A** grass
    - **B** snake
    - **C** fish
    - **D** raccoon

26. The following food chain is one food chain in the estuary ecosystem:

    algae → fish → raccoon → snake

    Which organism in the food chain is at the highest trophic level?
    - **A** algae
    - **B** fish
    - **C** raccoon
    - **D** snake

27. How does the fossil record show that species have changed over time?
    - **A** Features exist in extinct species that do not exist in modern species.
    - **B** Transitional species have features that exist in both extinct and modern species.
    - **C** Extinct species had traits that disappeared and then reappeared in modern species.
    - **D** Some transitional species show traits that are a combination of both extinct and modern species.

See page 174 for answers.

Lesson 1

1. cell membrane
2. mitochondrion
3. chloroplast
4. Golgi apparatus
5. nucleus, organelles with membranes
   Prokaryotic cells do not have a nucleus or organelles with membranes.
6. B. Plant cells have cell walls and chloroplasts.

Lesson 2

1. B.
2. B.
3. B. Replication occurs when the chromosomes copy themselves, as shown in the first two parts of the image.
4. B.
5. C. Because the new cells have the same genetic material as the parent cell, they are identical.
6. D. Reproductive cells, or gametes, are a result of the process of meiosis, not mitosis.
7. C. Transcription is the process by which a strand of RNA is created using the sequence of bases present in a particular strand of DNA.
8. D. Translation is the process by which transfer RNA brings the appropriate amino acids to the ribosomes so they can be bonded together to form a protein.

Lesson 3

1. A. Moss is a nonvascular plant, which means that it does not have vessels through which it can transport materials, such as nutrients.
2. A. Plant reproduction occurs in the flowers of the plant because that is where the plant's gametes, or reproductive cells, are found.
3. B.
4. D.
5. D. Xylem transports water from the roots to the leaves of a plant, so it extends from the roots to the leaves of a plant.
6. A. The bee moved the male gamete to the female reproductive organ.
7. A.

Lesson 4

1. D.
2. C. The end product of cellular respiration is water, which cannot be formed without oxygen.
3. B. If the stomata are closed, the grass cannot take in carbon dioxide from the air and will not be able to carry out photosynthesis.
4. Possible answer: Photosynthesis produces oxygen, which is needed for animal respiration. It also produces glucose, which is needed for animals to carry out cellular respiration.
5. Possible answer: Cellular respiration, which takes place in the mitochondria, results in the production of a large amount of ATP, which provides energy for the cell to carry on life functions.

Lesson 5

1. D. Reptiles, such as iguanas, are cold blooded, lay eggs, have scales, and live only on land.
2. D. Mammals, such as dogs, have a dorsal nerve cord, are warm blooded, give birth to live offspring, and have hair.
3. C. Amphibians and fish are both cold blooded. Fish have scales, and amphibians have smooth, moist skin.
4. D. You move your hand away from a hot stove without even thinking about it—this is because the message goes to your spinal cord and back to the muscle before the message is sent to your brain.
5. D. Sponges are in the group Porifera, They have no symmetry and no real tissues or organs.
6. C. Bees dance to show the location of food; this behavior is not taught but is something they do by instinct.
7. B. Snakes are reptiles and have scales. Toads are amphibians and have smooth, moist skin.
8. B. Both mammals and birds are warm blooded. Only mammals have hair and give birth to live young. Neither birds nor mammals have an exoskeleton.

Lesson 6

1. **A.** The products of digestion are transported where they need to go by the circulatory system.

2. **C.** The excretory system filters waste from the blood and allows the body to eliminate it as urine.

3. **C.** The muscles of the upper arm are under a person's control, so they are voluntary muscles.

4. **A.** The nervous and endocrine systems regulate the body's conditions.

5. **B.** The excretory system is responsible for the removal of metabolic waste in the form of urine.

6. **A.** The brain receives information about the outside environment through the senses, interprets that information, and then sends a signal to tell the body to perform an appropriate reaction.

7. **A.** Digestive enzymes are present in the mouth, stomach, and small intestine.

8. **A.** If the metabolic activities slow due to an underactive thyroid, a person may have difficulty maintaining body temperature and may not be able to conduct the activities required for the release of energy at an appropriate speed, resulting in a tired feeling.

9. **C.** Enzymes are present in the stomach, where they help digestion occur.

10. **C.** The respiratory system brings in oxygen, which is used to release the energy stored in the food we eat.

11. **C.** Capillaries are tiny blood vessels through which materials move into and out of the cells of the body.

12. **D.** Information from the environment comes in through your ears and goes to the brain via nerves, where it is interpreted; a reaction message goes to the spinal cord via nerves and then to the muscles, again via nerves.

Lesson 7

1. **D.** A virus is an infectious pathogen because it is a disease-causing organism.

2. **B.** Vaccines contain weak or dead pathogens that are unable to infect the body.

3. **C.** A hereditary disease is passed on from parents to offspring.

4. **D.** Mucus membranes are part of a person's innate immunity, because a person is born with their mucus membranes, and they are the first line of defense against infection.

5. **A.** Acquired immunity develops over time. As you are exposed to pathogens, your body builds up a memory to those pathogens in order to defeat them quickly.

6. **A.** Inflammation allows an area to be cordoned off so that the infection has less chance of spreading.

7. **A.** If a person has an immunodeficiency, his ability to fight infection is weakened.

8. **D.** Coughing prevents pathogens from reaching your lungs because it forces the pathogen out of your respiratory system; this is an ability people are born with.

Lesson 8

1. **C.**

2. **A.**

3. **B. or D.**

4. **B. or D.** Answers B and D, the two bird species, are the most closely related species. They belong on the two branches of the cladogram separated by just one branching event. The two branches are equivalent, so either answer could go on either line 3 or 4.

5. **Domain**

6. **Phylum**

7. **Order**

8. **Family**

9. **Genus**

10. **Homo sapiens** The scientific name consists of the genus and species names together.

11. **C.** Taxonomists try to classify species based on evolutionary relationships. How long ago did two species have a common ancestor? DNA evidence is a good way to learn about these relationships.

12. **Sample answer: A species is a group of organisms that can interbreed and produce offspring that can interbreed.**

Lesson 9

1. DNA
2. chromosomes
3. mutations
4. phenotype
5. alleles
6. genotype
7. pink If the pink allele is dominant, some of the pink parents may have the white allele. When two pink parents with the white allele reproduce, 25% of their offspring will be white. If the white allele is recessive, none of the white parents will have the pink allele, so none of their offspring will have the pink allele.
8. B. Most changes to an organism's body cannot be inherited. In humans, only genetic mutations that affect sperm or egg cells can be passed down to children. Some types of pollution and radiation can make genetic mutations more common.
9. C. All four have the same eye color, which means that they have the same phenotype. Identical twins always have the same genotype for every trait, since their DNA is identical. However, the fraternal twins do not necessarily have the same genotype as the identical twins, or even the same genotype as each other.
10. A. If the two alleles were equally common, the phenotype associated with the dominant allele (dimples) would be more common than the other phenotype. Since dimples are not the most common phenotype in the population, the allele for no dimples must be more common.
11. AO or OA
12. OO
13. AO or OA
14. OO
15. ½ Boxes 11 and 13 will be AO, like their mother.
16. type O
17. ½ Boxes 2 and 4 will have a genotype OO and a phenotype of type O blood, like their father.

Lesson 10

1. a trait that serves some kind of an improved function within a population
2. the process of nature selecting for an advantageous trait/adaptation within a population
3. a species from which all members of a population descended
4. Possible answer: Without variation, adaptations that are disadvantageous will cause a species to die out. Natural selection cannot occur if a population has no variation of traits in it.
5. C. Natural selection occurs when adaptations that are advantageous are passed to subsequent generations. After a period of time, all members of the species will possess that adaptation.

Lesson 11

1. biotic
2. biotic
3. abiotic
4. abiotic
5. Possible answer: The trees being cut down can disrupt the ecosystem because some animals can lose their home. In addition, the trees will no longer be able to take carbon out of the ecosystem. The factory will add carbon into the ecosystem. This will result in a disruption of nutrient cycling in this ecosystem.
6. D. A biome can sometimes be thought of as an ecosystem on a very large scale.

Lesson 12

1. Possible answer: An omnivore can be a primary and secondary consumer. For example, when eating plants, the omnivore is considered a primary consumer, but if the animal eats another animal, then it is a secondary consumer.
2. B. A food chain shows only one path leading to the top predator. Any of the organisms in the food chain can also belong to another food chain.
3. C. A secondary consumer eats a primary consumer. The primary consumer is the zooplankton, which makes the fish a secondary consumer.
4. B. Energy is lost as the food chain goes up in trophic level; therefore, the highest level of the food chain has the least amount of energy.

Unit Test

1. C.
2. B.
3. D.
4. A.
5. C.
6. C.
7. D.
8. D.
9. A.
10. D.
11. A.
12. D.
13. B.
14. A.
15. D.
16. B.
17. D.
18. B.
19. D.
20. A.
21. C.
22. A.
23. D.
24. D.
25. A.
26. D.
27. A.

- **abiotic component** – the non-living factors in an ecosystem
- **active transport** – the movement of molecules from an area of high concentration to low concentration; this movement requires energy
- **adaptation** – a trait that is common in a population because it provides an improved function
- **allele** – one of multiple versions of the same gene
- **amino acid** – a molecule that is the building block of proteins
- **ATP** – adenosine triphosphate; a molecule that carries energy
- **behavior** – the way in which one acts
- **biome** – a community existing of many ecosystems
- **biotic component** – the living factors in an ecosystem
- **bird** – vertebrate animal that has feathers
- **carbon dioxide** – a molecule that is a product of cellular respiration and a reactant in photosynthesis
- **cell membrane** – the outer layer of an animal, fungus, bacteria, or protist cell, which determines what enters or leaves the cell
- **cell wall** – a tough outer layer that surrounds the cell membrane in some types of cells, including plant cells but not animal cells
- **cellular respiration** – a process that transfers the energy stored in glucose into molecules of ATP
- **chlorophyll** – a substance within the chloroplasts of cells that gives plants their green color and absorbs light energy from the sun
- **chloroplast** – an organelle found in plant cells and eukaryotic algae that captures the energy from sunlight and makes it into food for the cell
- **chromosome** – a linear or circular strand of DNA that contains the genetic material of a cell, sometimes bonded to proteins
- **circulatory system** – the body system responsible for the transport of materials
- **cladogram** – a branching diagram that shows evolutionary relationships among organisms
- **cold blooded** – a condition whereby the internal temperature of an animal's body varies with that of the environment
- **common ancestor** – an organism from which other species have descended
- **cytoplasm** – the contents of a cell within the cell membrane and outside of the nucleus
- **decomposers** – organisms that break down organic matter
- **diffusion** – the process by which molecules mix as a result of their movement
- **digestive system** – body system responsible for the breakdown of food
- **disease** – a condition that negatively affects the body
- **diversity** – variation of living things
- **DNA** – deoxyribonucleic acid, the molecule that carries the genetic instructions that living things use to develop, function, and reproduce
- **domain** – the broadest grouping of organisms used by many taxonomists; can contain one or more kingdoms
- **ecosystem** – a particular environment and all its living and non-living components
- **endocrine system** – body system, made up of glands, that regulates conditions in the body
- **endoplasmic reticulum** – a system of interconnected, folded sacs and tubes that produces materials, especially proteins, and transports them through the cell
- **environmental diseases** – diseases that are the result of exposure to certain environmental conditions; for example, lead poisoning
- **enzymes** – proteins that speed up chemical reactions but do not change themselves

- **esophagus** – muscular tube through which food passes; extends from the mouth to the stomach
- **eukaryote** – an organism whose cells contain a nucleus and other organelles that are contained within membranes; member of the domain Eukaryota
- **evolution** – genetic change over time
- **excretory system** – body system responsible for the removal of metabolic waste
- **fertilization** – the union of male and female gametes to produce a new organism
- **food chain** – the sequence in which matter and energy are transferred in an ecosystem
- **food web** – a diagram that shows interconnections of all species in an ecosystem
- **fossil record** – the use of fossils in sedimentary rock layers to describe the past
- **gametes** – a reproductive cell that unites with another of the opposite sex to form offspring
- **gene** – a section of DNA located at a specific point on the chromosome that determines a characteristic of an organism by providing instructions for building proteins
- **genetics** – the study of how genes work and how they are transferred from parents to offspring
- **genotype** – the particular alleles carried by an individual organism
- **germinate** – the process by which a plant begins to grow from a seed
- **Golgi apparatus** – an organelle that packages and sorts molecules for transport through the cell
- **heredity** – the genetic transmission of characteristics from parent to offspring
- **immunity** – the body's ability to fight off disease
- **immunization** – the process by which someone is made immune to a disease, generally through the use of a vaccine
- **immunodeficiency** – the failure of the immune system to protect the body from disease
- **inflamation** – swelling
- **instinct** – a complex behavior that an animal is born knowing how to perform in response to a specific stimulus
- **kingdom** – a taxonomic rank below domain that can contain one or more phyla; used by some taxonomists
- **leaves** – the part of the plant where photosynthesis occurs
- **mammal** – a warm-blooded vertebrate with hair that gives birth to live young; example, rabbit
- **meiosis** – cell division that results in the formation of gametes
- **messenger RNA** – a type of RNA molecule that carries the DNA message, or information
- **metabolism** – interrelated chemical reactions within cells that allow organisms to carry out life processes
- **minerals** – a class of substances occurring in nature
- **mitochondrion** – a structure that provides energy to the cell by converting food into ATP, the cell's energy source
- **mitosis** – cell division that results in the formation of new, identical cells
- **muscular system** – the body system responsible for movement
- **mutation** – a permanent change in the DNA sequence of an organism, usually caused by an error when DNA is copied
- **natural selection** – the theory that nature will select for traits that are advantageous to the species
- **nervous system** – the system of the body that regulates and responds to conditions inside and outside the body
- **nonvascular plant** – a type of plant that does not contain real roots, stems, or leaves, or vascular tissue

- **nucleus** – the control center of a eukaryotic cell; contains the genetic material (DNA) that directs the activities of the cell
- **nutrients** – substances that provide nourishment
- **nutrition** – the process by which living things get and process nutrients
- **order** – a taxonomic rank below class that can contain one or more families
- **organelle** – membrane-bound structures within eukaryotic cells that carry out specific functions for the cell
- **osmosis** – the process by which water moves across the cell membrane from an area of high concentration to one of low concentration
- **oxygen** – a molecule that is a product of photosynthesis and a reactant in cellular respiration
- **passive transport** – the process by which molecules move from an area of high concentration to low concentration; no energy is required for this process
- **pathogen** – a disease-causing organism
- **phenotype** – observable characteristic(s) of an organism that are expressed as a result of its genotype
- **phloem vessel** – a tissue found in plants responsible for transporting sugar and other products of metabolism
- **photosynthesis** – a process in which plants convert light energy to chemical energy
- **pistil** – the female reproductive organs of a plant
- **pollen** – a powdery substance that contains the male gametes of a plant
- **pollination** – the process in plants in which pollen is moved from the anther to the pistil
- **prokaryote** – a single-celled organism that does not contain a membrane-bound nucleus or other organelles and that is a member of either the domain Archaea or the domain Bacteria
- **protein** – substance that carries out cell functions such as breaking down nutrients
- **reflex** – an action that is performed without conscious thought in response to a stimulus
- **replication** – process in which the DNA of chromosomes is copied
- **respiratory system** – the body system that controls gas exchange, including breathing
- **ribosome** – a cell's organelle where protein synthesis takes place
- **reptile** – cold-blooded vertebrate that has scaly skin and lays eggs; example, turtle
- **RNA** – ribose nucleic acid, a molecule that carries the genetic code from the nucleus to the cytoplasm of a cell
- **roots** – the part of the plant that attaches to the ground and absorbs water and minerals from the soil
- **seed** – the object produced from the process of plant fertilization, from which plants grow
- **sense organs** – organs of the nervous system that take in information from the environment; for example, the eyes and nose
- **skeletal system** – the body system comprised of bones; it is responsible for movement, support, and protection of vital organs
- **species** – a group of organisms that can interbreed
- **stamen** – the male reproductive organ found within a flower
- **stem** – the main stalk of a plant
- **symbiosis** – a relationship between two different species that have a close physical interaction with one another
- **symmetry** – having identical parts or characteristics on all sides
- **taxonomy** – the science of naming and classifying living organisms
- **traits** – features

Unit Glossary

- **transcription** – the process by which the information from a strand of DNA is copied into a new molecule of RNA
- **transfer RNA** – a type of RNA molecule that carries amino acids
- **translation** – the process by which the genetic code is decoded to produce a sequence of amino acids for a specific protein
- **trophic level** – the position an organism occupies in a food chain
- **urine** – a liquid comprised of the wastes of the body's metabolic reactions
- **vascular plant** – a type of plant that has vascular tissue or vessels, as well as true roots, leaves, and a stem
- **voluntary muscles** – muscles that are under conscious control
- **warm blooded** – a condition whereby an animal's internal temperature is kept constant
- **white blood cells** – blood cells that fight infection
- **xylem** – tissue found in vascular plants that transports water and minerals from the roots to other parts of the plant

UNIT 4

Study More!

Consider exploring these concepts, which were not introduced in the unit:

- Lysosome, nucleolus, ribosome, vacuole
- Phases of mitosis and meiosis: prophase, metaphase, anaphase, telophase
- Haploid and diploid
- Carotene
- Gametophytes
- Nodules
- Spores
- Accessory pigments for photosynthesis
- Axons, dendrites, synapses
- Cardiac muscles, valves, ventricles
- Joints, ligaments, tendons
- Reproductive system
- Spinal cord
- Bone marrow
- Blood platelets
- Liver and gall bladder
- Health regulations and public health
- Medical defenses
- Fraternal and identical twins
- Cloning
- Evolutionary fitness
- Details of the carbon and nitrogen cycles
- Calories

Answer the following questions based on the material you have learned.

Use the following passage to answer questions 1–3.

Calcium has an atomic number of 20 and is in the second column from the left on the periodic table. Nitrogen has an atomic number of 7 and is in the fourth column from the right on the periodic table. When calcium and nitrogen react, they form calcium nitride, with the formula Ca_3N_2.

1. What physical properties is calcium expected to have?
   A dull, brittle, poor electricity conductor
   B shiny, brittle, good electricity conductor
   C dull, malleable, poor electricity conductor
   D shiny, malleable, good electricity conductor

2. How many electrons are in a neutral nitrogen atom?
   A 4
   B 7
   C 11
   D 14

3. Which best describes the nature of the bonds in calcium nitride?
   A Calcium and nitrogen share six electrons in a triple covalent bond.
   B Calcium and nitrogen share two electrons in a single covalent bond.
   C The positively charged calcium ions are attracted to the negatively charged nitride ions.
   D The negatively charged calcium ions are attracted to the positively charged nitride ions.

4. Which of these is an example of erosion?
   A Acid rain reacts with a rock and gradually weakens it.
   B Waves beating on the seashore turn rocks into sand.
   C Water freezing and thawing creates a large crack in a rock.
   D A flood washes away sand from a river bank.

Use the following image to answer questions 5 and 6.

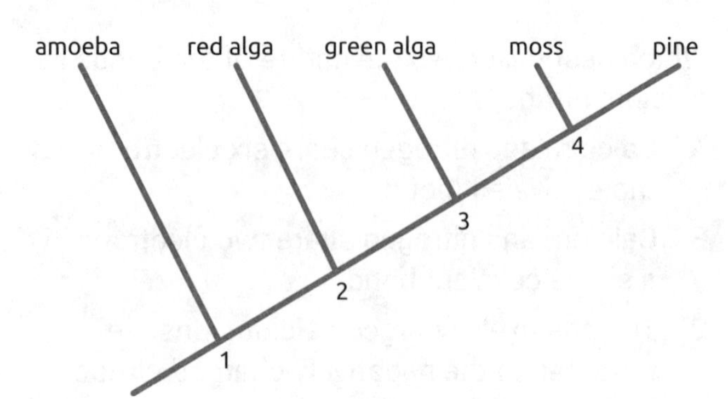

5. A common ancestor of all organisms can be found at which point?
   A 1
   B 2
   C 3
   D 4

6. Which organisms are most closely related?
   A moss and pine
   B amoeba and pine
   C pine and green alga
   D green alga and moss

7. When it is summer in the Northern Hemisphere, which of the following is NOT true?
   A The Northern Hemisphere is tilted toward the sun.
   B The Northern Hemisphere is closer to the sun than at other times of year.
   C The Northern Hemisphere receives more direct sunlight than the Southern Hemisphere.
   D The Northern Hemisphere experiences longer days than the Southern Hemisphere.

8. Water has the formula H_2O. In the molecule, the oxygen end of the molecule is slightly negative, and the other end is slightly positive. Which is the best description of the water molecule?
   A ionic
   B polar
   C metallic
   D non-polar

9. What causes tectonic plates to move?
   A the pull of the moon's gravity
   B Earth's rotation
   C pressure from earthquakes
   D movement of hot magma under the plates

Use the following passage to answer questions 10–12.

A chemistry student mixed sulfur and iron in a test tube and then briefly heated the mixture over a flame, while monitoring the temperature inside the test tube. When the mixture began to glow, the student removed the test tube from the flame. The mixture continued to glow, and the test tube temperature continued to increase for several minutes. At that point, the glow faded away, and the mixture turned black. The student waited until the test tube cooled and then tested the black material, which turned out to be iron sulfide (FeS). The balanced equation for the reaction is $Fe(s) + S(s) \rightarrow FeS(s)$.

10. What type of reaction occurred in the experiment?
    A synthesis
    B decomposition
    C single displacement
    D double displacement

11. The student combined equal amounts of iron and sulfur and found that all of it reacted to form iron sulfide. What would happen if the student doubled the amount of iron?
    A The amount of iron sulfide produced would double.
    B The amount of iron sulfide produced would decrease.
    C The amount of iron sulfide produced would stay the same.
    D The amount of iron sulfide produced would increase by 50%.

12. How should the student explain the behavior of the materials during the reaction?
    A The reaction is exothermic, because it released more heat than it absorbed.
    B The reaction is exothermic, because it absorbed more heat than it released.
    C The reaction is endothermic, because it released more heat than it absorbed.
    D The reaction is endothermic, because it absorbed more heat than it released.

13. Which of the following most likely came about due to natural selection?
    A a scar on a child's knee
    B a peacock's colorful feathers
    C the curiosity of a young chimpanzee
    D the strong arm muscles of a baseball pitcher

14. Which of the following actions must take place in order for a seed to germinate?
    A The seed must be fertilized.
    B The seed must be pollinated.
    C The seed must get the right amount of water.
    D The seed must get the right amount of sunlight.

Use the following passage to answer questions 15–17.

A pair of circus performers is practicing a stunt for the show. When Waldo jumps off the platform and lands on the end of the lever, his end of the lever moves to the ground and Zelda flies up into the air. Waldo's mass is 80 kg, and Zelda's mass is 60 kg.

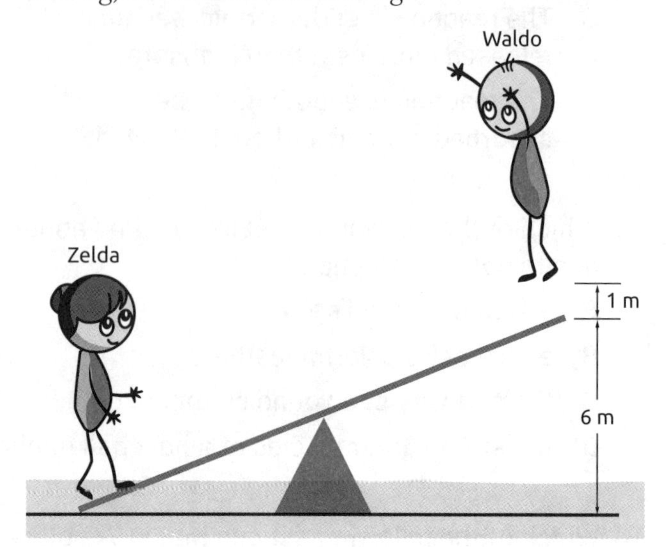

15. What is Waldo's potential energy when he is on the platform?
    - **A** 49 J
    - **B** 131 J
    - **C** 3,920 J
    - **D** 4,704 J

16. When Waldo jumps, his acceleration toward the ground is 9.8 m/s^2. What is his velocity downward after 0.4 seconds?
    - **A** 1.6 m/s
    - **B** 3.9 m/s
    - **C** 9.8 m/s
    - **D** 24.5 m/s

17. When Waldo and the lever hit the ground, his velocity is 9.4 m/s downward. What is Zelda's velocity upward?
    - **A** 4.7 m/s
    - **B** 9.4 m/s
    - **C** 12.5 m/s
    - **D** 18.8 m/s

18. Which animal has a segmented body and an exoskeleton?
    - **A** ant
    - **B** clam
    - **C** earthworm
    - **D** sponge

The North Atlantic Gyre is a system of ocean currents that circulate water around the northern Atlantic

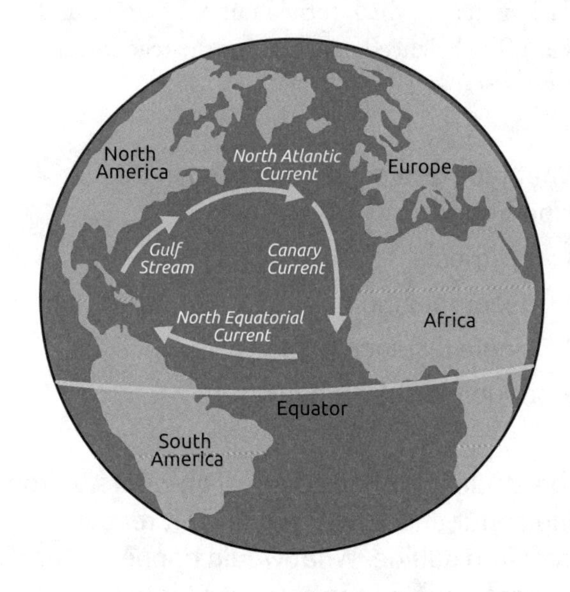

Ocean, as shown in the following diagram.

19. Which statement best describes how the temperatures of the Gulf Stream and the Canary Current compare?
    - **A** The Gulf Stream is warmer because it carries warm water from nearer the equator, where the oceans receive more solar radiation.
    - **B** The Canary Current is warmer because it carries warm water from southern Europe, which is sunnier than North America.
    - **C** Both currents are the same temperature because the same water moves in a giant circle around the Atlantic.
    - **D** The Canary Current is warmer because it carries warm water from farther north in the Atlantic, where the oceans receive more solar radiation.

20. Which describes the difference between a food web and a food chain?
   - **A** A food web shows the energy flow within an ecosystem.
   - **B** A food web shows the biotic and abiotic factors in an ecosystem.
   - **C** A food chain shows all of the interactions between organisms within an ecosystem.
   - **D** A food chain shows the evolutionary relationships between organisms in an ecosystem.

Use the following passage to answer questions 21 and 22.

In humans, the presence of freckles (F) is dominant to having no freckles (f). A man with freckles and his wife, who has no freckles, have 8 children. Of those children, 4 have freckles, and 4 have no freckles.

21. What are the most likely alleles of the parents?
   - **A** Ff and Ff
   - **B** Ff and ff
   - **C** FF and ff
   - **D** FF and Ff

22. If one of the offspring who has freckles has children with a person who has the same genotype, what will be the most likely phenotypes of the children?
   - **A** 25% freckled, 75% non-freckled
   - **B** 50% freckled, 50% non-freckled
   - **C** 75% freckled, 25% non-freckled
   - **D** 100% freckled, 0% non-freckled

23. A kidney infection may make which of these activities painful?
   - **A** blinking
   - **B** breathing
   - **C** eating
   - **D** urinating

24. Which type of energy transfer can occur in a vacuum?
   - **A** radiation
   - **B** conduction
   - **C** convection
   - **D** electric current

25. Which of the following is a way that human activity can increase the greenhouse effect, causing Earth's temperature to rise?
   - **A** Using nitrogen fertilizers pollutes water, harming aquatic ecosystems.
   - **B** Air pollution damages the ozone layer, allowing more ultraviolet radiation to reach Earth's surface.
   - **C** Installing solar panels can reflect too much solar radiation into the atmosphere.
   - **D** Factory farming of livestock adds methane to the atmosphere.

Use the following image to answer questions 26–28.

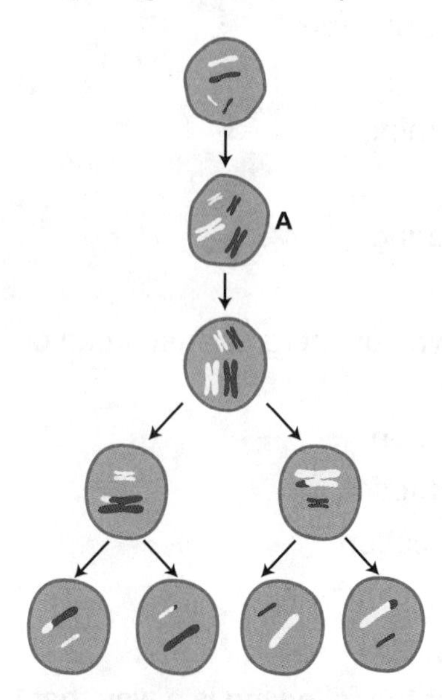

26. The cell division shown in the image results in which type of cells?
    - **A** blood
    - **B** brain
    - **C** gametes
    - **D** muscle

27. Which is true for the final four cells?
    - **A** They each have a set of chromosomes identical to the parent cell.
    - **B** They each have half the number of chromosomes as the parent cell.
    - **C** They each have twice the number of chromosomes as the parent cell.
    - **D** They each have a set of chromosomes identical to the other resulting cells.

28. What is occurring at A?
    - **A** Chromosomes are splitting.
    - **B** Chromosomes are replicating.
    - **C** Chromosomes are translating.
    - **D** Chromosomes are transcribing.

29. A student measures the pH of solution X and determines the pH to be 9. The student adds a few drops of solution Y and determines the pH of the new solution to be 7. How should the student describe the two solutions?
    - **A** Solution X is a base, and solution Y is an acid.
    - **B** Solution X is a base, and solution Y is neutral.
    - **C** Solution X is an acid, and solution Y is a base.
    - **D** Solution X is an acid, and solution Y is neutral.

30. Cells in the leg muscles of a runner would most likely have more of which organelle?
    - **A** cytoplasm
    - **B** endoplasmic reticulum
    - **C** Golgi apparatus
    - **D** mitochondrion

Use the following passage and table to answer questions 31 and 32.

A student is testing the rate of photosynthesis in plants. She knows that when carbon dioxide is placed in a solution of water, the solution of water becomes slightly acidic. When the carbon dioxide is removed, the pH goes back up. She puts tap water into three test tubes and tests the pH. The pH is slightly basic, with a pH of 7.4.

In the first setup, she uses a straw to breathe into a test tube with water.

In the second setup, she places a plant, elodea, into a test tube with water after wrapping the plant in aluminum foil.

In the third setup, she places the plant alone into the test tube with water.

Her results are shown in the table.

Setup	1: No plant in water, only breath	2: Plant with aluminum foil in water	3: Plant alone in water
pH Results	Water is acidic.	Water is acidic.	Water is slightly basic.

31. Which best explains why the water in Setup 3 is slightly basic?

    A Photosynthesis is occurring, producing carbon dioxide and making the solution less acidic.

    B Respiration is occurring, producing carbon dioxide and making the solution less basic.

    C Photosynthesis is occurring, taking up some of the carbon dioxide and making the solution less acidic.

    D Respiration is occurring, taking up some of the carbon dioxide and making the solution less basic.

32. Which best explains the results in Setup 2?

    A Because the plant is covered, it is dying; as a result, it releases oxygen into the solution.

    B Because the plant is covered, it respires at a faster rate, releasing large amounts of oxygen into the solution.

    C Because the plant is covered, it is able to use the products of respiration for photosynthesis, producing carbon dioxide.

    D Because the plant is covered, is it unable to conduct photosynthesis, but respiration is occurring, producing carbon dioxide.

33. Which characteristic is shared by all vertebrates?
    - **A** warm blooded
    - **B** radial symmetry
    - **C** no dorsal nerve cord
    - **D** internal body skeleton

34. What role do fossil fuels play in the carbon cycle?
    - **A** Carbon from the atmosphere is deposited into fossil fuels, and then it enters plants and animals when they take it up from the soil.
    - **B** Carbon in fossil fuels comes from bacteria in the soil, and then it enters the atmosphere when fuels are burned.
    - **C** Carbon from the atmosphere is deposited into fossil fuels, and then it enters the soil when it rains.
    - **D** Carbon in fossil fuels comes from dead plants and animals, and then it enters the atmosphere when fuels are burned.

Use the following image to answer questions 35 and 36.

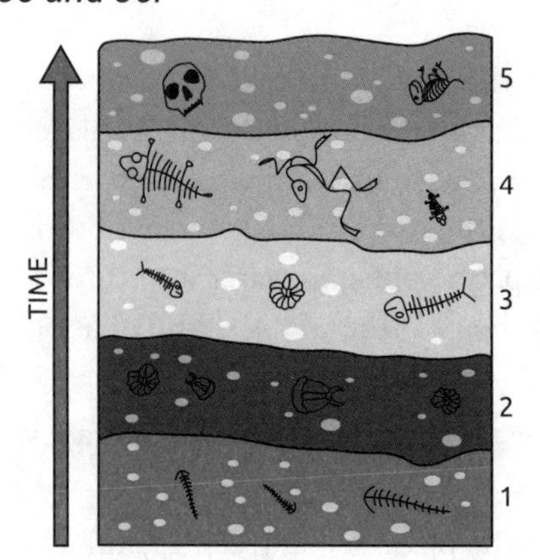

35. Which layer contains fossils of organisms that evolved most recently?
    - **A** 2
    - **B** 3
    - **C** 4
    - **D** 5

36. In which layer is the common ancestor of the fossils in layers 2 and 3 most likely found?
    - **A** 1
    - **B** 2
    - **C** 3
    - **D** 4

37. In which structures of the respiratory system does gas exchange occur?
    - **A** alveoli
    - **B** arteries
    - **C** bronchi
    - **D** capillaries

38. Lucinda wants to take her friend to see a shipwreck in the bay near her house. The shipwreck is usually under water, but during spring tides, part of the ship is visible from shore at low tide. If it is winter and the moon was full five days ago, when is the next time Lucinda can take her friend to see the shipwreck?
    - **A** in a few days, at low tide during the third quarter moon
    - **B** next week, at low tide during the next new moon
    - **C** next month, at low tide during the next full moon
    - **D** in a few months, at the first low tide during spring

39. A substance that has no definite shape or volume undergoes a phase change, after which it still has no definite shape but now has a constant volume. Which phase change did the substance experience?
    - **A** boiling
    - **B** melting
    - **C** freezing
    - **D** condensing

40. In order to produce a seed, which of the following processes is necessary?
    - **A** Pollen must be taken from the anther to the ovule, where fertilization occurs.
    - **B** Pollen must be taken from the pistil to the anther, where germination occurs.
    - **C** Pollen must be taken from the pistil to the anther, where fertilization occurs.
    - **D** Pollen must be taken from the anther to the ovule, where germination occurs.

Use the following passage to answer questions 41 and 42.

Erythropoietin is a hormone produced by the kidneys in response to deficient oxygen levels. Low oxygen levels can be caused by a number of factors, including blood loss, lung disease, or high altitudes (where less oxygen is available).

Erythropoietin increases the production of red blood cells by the bone marrow. When the proper levels of oxygen are restored, the production of erythropoietin decreases, and red blood cell production returns to normal.

41. In the scenario described, which systems are working together?
   - **A** endocrine, circulatory, and respiratory
   - **B** digestive, circulatory, and respiratory
   - **C** endocrine, digestive, and excretory
   - **D** digestive, circulatory, and endocrine

42. Which term describes the way the body has maintained proper levels of oxygen?
   - **A** innate behavior
   - **B** regulation
   - **C** reflex
   - **D** metabolism

43. Evidence suggests that the universe is currently changing in which of these ways?
   - **A** It is becoming larger and hotter.
   - **B** It is becoming larger and cooler.
   - **C** It is becoming smaller and hotter.
   - **D** It is becoming smaller and cooler.

44. The 2010 earthquake that devastated Haiti had a magnitude of 7.0. Four years later, an earthquake of magnitude 6.0 struck the Napa Valley of California. How much bigger was the Haiti earthquake, compared to the Napa Valley one?
   - **A** $1\frac{1}{6}$ times bigger
   - **B** 2 times bigger
   - **C** 10 times bigger
   - **D** 100 times bigger

Use the following passage to answer questions 45 and 46.

Measles is a disease caused by a virus. Symptoms include a fever, rash, dry cough, and runny nose. In 1963, a vaccine for the measles became available. The vaccine is now provided to very young children.

45. What type of disease is measles?
   - **A** age related
   - **B** environmental
   - **C** genetic
   - **D** infectious

46. The measles vaccine allows people to
   - **A** access a cure for measles.
   - **B** develop an active immunity to measles.
   - **C** develop an innate immunity to measles.
   - **D** access medicine to treat the symptoms of measles.

47. In which of the following situations is the most work being done?
    A A 10 kg rock is accelerated at 15 m/s^2 for 10 meters.
    B A 20 kg rock is accelerated at 15 m/s^2 for 10 meters.
    C A 20 kg rock is pushed with a force of 90 N for 10 meters.
    D A 10 kg rock is pushed with a force of 90 N for 10 meters.

48. When heat and pressure under the earth's surface cause a rock to be transformed and hardened, what is the resulting rock called?
    A igneous
    B metamorphic
    C granitic
    D sedimentary

49. Which is part of one's adaptive immunity?
    A skin
    B pathogens
    C white blood cells
    D mucus membranes

50. Natural processes are constantly breaking rocks down and carrying them away. When sediments worn from old rocks are carried away, this is called erosion. What is the opposite process called, when these materials are used to create a new landform?
    A chemical weathering
    B deposition
    C physical weathering
    D decomposition

See page 190 for answers.

Practice Test Answer Key

1. D.
2. B.
3. C.
4. D.
5. A.
6. A.
7. B.
8. B.
9. D.
10. A.
11. C.
12. A.
13. B.
14. C.
15. D.
16. B.
17. C.
18. A.
19. A.
20. A.
21. B.
22. C.
23. D.
24. A.
25. D.
26. C.
27. B.
28. B.
29. A.
30. D.
31. C.
32. A.
33. D.
34. D.
35. D.
36. A.
37. A.
38. B.
39. D.
40. A.
41. A.
42. B.
43. B.
44. C.
45. D.
46. B.
47. B.
48. B.
49. C.
50. B.

NOTES

NOTES